THE INEVITABLE AI

ART OF GROWTH WITH GENERATIVE INTELLIGENCE

DR. ASWIN CHANDARR

STARDOM BOOKS

www.StardomBooks.com

STARDOM BOOKS
A Division of Stardom Publishing
and infoYOGIS Technologies.
105-501 Silverside Road
Wilmington, DE 19809

FIRST EDITION MARCH 2024

STARDOM BOOKS

A Division of Stardom Alliance
105-501 Silverside Road Wilmington, DE 19809,
USA

www.stardombooks.com

Stardom Books, United States
Stardom Books, India

THE INEVITABLE AI

Dr. Aswin Chandarr

p. 356
cm. 13.5 X 21.5

Category: COM004000 COMPUTERS/Artificial
Intelligence/General

ISBN: 978-1-957456-50-8

DEDICATION

To little Ayra, who inspires me to envision the future, and to my partner Rev, who grounds me in the beauty of the present —both the heartbeats of my life's work.

.

CONTENTS

ACKNOWLEDGMENTS

This book is not just a culmination of my thoughts and insights but a mosaic of inspiration, support, and wisdom from numerous remarkable individuals whose influence has been pivotal in its creation.

First and foremost, my gratitude extends to U. Mahesh Prabhu, who planted the initial seed for this journey. His vision was the spark that ignited my passion for writing this book. My daughter Ayra, with her boundless love and vibrant energy, reminds me daily of the brighter future we owe to the next generations.

My tenure at Loop Robots was a chapter of my career and a pivotal turn in my journey. My colleagues and friends enriched me, such as Mitsos, Kim, Maarten, Per, Mark, Job, Guus, Cezary, Mathijs, and others. The turn of events during this period invariably created time and space for this work to come to fruition.

To Prof. Pieter Jonker and Maja Rudinac, who taught me the art of simplifying the complex and applying first principles thinking, your mentorship has been invaluable in demystifying the technical world and focusing on purpose-driven real-world applications.

In retrospect, Martin Roos illuminated the path to reconciling grand aspirations with challenging realities. This lesson has been instrumental in shaping this work.

To my esteemed professors and lecturers at NIT-Trichy, particularly Dr. Sundareswaran and Dr. Sankaranarayanan, who ignited and nurtured my passion for engineering. They guided me through the formative stages of my academic journey, encouraging me to explore the profound questions that drive innovation and progress in our world.

To my mother, Raji, who sacrificed the prime of her life to provide me with an enriching childhood and patiently answer my endless questions of "WHY?" And to my father, Balu, the silent force behind my endeavors, offering unwavering support throughout my journey.

My in-laws' understated yet significant support, Prema and Nagarajan, has strengthened and encouraged me.

This is a nod to my great-great-grandfather, Alagarsamy, whose foresight and endeavors dating back to 1903 leave an indelible mark on our lives today.

To my circle of friends - Karthik, Rupa, Arun, Mukunda, and many more - your contributions, no matter how small, have left indelible marks on this work.

A heartfelt thank you to the editorial team at Stardom, especially Mr. Raam, Ranjitha, Rekha, and Priya, who meticulously sculpted the raw manuscript into the polished gem it is today.

Sergio, your philosophical musings and vision have been a cornerstone in shaping the content and soul of this book.

Trace Harris, our serendipitous meeting on the shores of Zanzibar was the genesis of a title that truly encapsulates the essence of this work.

With its majestic bridges and tranquil waterfronts, the city of Rotterdam has been the backdrop against which many of this book's ideas were conceived and nurtured.

A special thanks to Angela, Leman, and the wonderful souls who ensured Ayra's happiness and well-being, allowing me the peace of mind to devote to this book.

And to the countless engineers, scientists, and entrepreneurs whose relentless pursuit of innovation in AI has paved the way for a future we are only beginning to imagine.

This book is a tribute to each of you whose influence has guided this journey.

Lastly, my gratitude will not be complete without a special note to my dear Rev, the unparalleled co-pilot of my life's journey, work endeavors, and soulful explorations. She's the one person who can engage in a spirited debate on the cataclysmic demise of supernovas one moment and ponder the existential dilemmas of life the next, all while managing not to roll her eyes at my latest, most incredible (and occasionally not-so-great) ideas.

Embarking on the entrepreneurial path is akin to mixing a complex cocktail of risk, adrenaline, sleepless nights, promises, and the wild highs and lows of creating something from nothing. A concoction that tests the sturdiest of relationships, especially when the boundaries of work and life blur into a continuous "mind always ON" mode.

This adventure could easily blur the lines between work and life, making "us time" feel more like "brainstorming session" time. Yet, Rev has stood unwaveringly by my side. Finding words to express my gratitude for Rev is like trying to describe the taste of water— indispensable but ineffably complex.

She steadfastly believes in the greener grass beneath our feet, even when our garden seemed more like a battlefield of unrealized ventures. Her unwavering support, a beacon of belief that outshines any doubt, makes her the extraordinary woman who not only ignites the joy of creation but also perfectly complements my left-brained tendencies, forming a whole far greater than the sum of its parts. Much like the best parts of my life, this book wouldn't be without her.

PREFACE

You've probably heard the term "AI" more times than you can count! It's everywhere - in the news, social media, and daily chats. Our social feeds teem with tales of AI's prowess: from ChatGPT's marketing marvels to mesmerizing AI-generated art which gives form to the creativity of keyboard artists, birthing memes and garnering viral shares, from deep fakes to false information.

But amidst this buzz, many share an underlying sentiment: What exactly is AI? And, more importantly, why should I care? If we introspect ourselves, for many of us, the true essence of Artificial Intelligence is still a bit of a mystery.

Picture this ancient tale: A group of blind men encounters an elephant. Each touches a different animal part and then describes the elephant based on that limited interaction. One feels the trunk and believes it's a snake. Another touches the leg, thinking it's a tree. That's how most of us are with AI. We hear bits and pieces, but the complete picture? It often needs to be included. We are all fascinated yet a bit intimidated.

Now, the burning question: Is AI just a fleeting fad? A temporary storm that'll pass? Or is it a transformative force poised to redefine our very existence? Given the leaps in technology, especially the strides in Generative AI, the evidence leans heavily towards the latter. AI is no longer just about making machines play chess or recommend songs. It's not just about chatbots or virtual assistants anymore. Poised to reshape every digital interaction we experience, its applications stretch far beyond the buzzworthy realms of copywriting and graphic creation. Its influence is bound to permeate sectors as varied as healthcare, education, finance, governance, and even niche areas like biotechnology and space exploration.

This book is your companion, guide, and map through the intricate maze of AI. We aim to simplify the complex and clarify where there's confusion. Whether debunking myths or highlighting opportunities, we've got you covered. And the best part? There is no need for a tech degree to understand. It's a promise. This is your friendly guide to the world of AI. It's like chatting with a knowledgeable friend over coffee, breaking down the ins and outs of this transformative technology.

AI isn't just about tech geeks and big companies. It's about all of us. Are you a professional watching colleagues whisper about the marvels of AI and feeling a tad lost? Will it be an asset or a looming threat to your job? Are you an entrepreneur hunting for the next big idea, scouting for innovative vistas? Are you an enterprise leader charting an AI-forward strategy? The potential applications of AI might be intriguing. Investors might be grappling with the dual sides of AI—the lucrative returns it promises and the inherent uncertainties it brings. Policymakers strive to strike a balance between fostering innovation and safeguarding societal interests. Or perhaps you're just a curious soul eager to keep up with the times and understand the technology shaping our world. Whoever you are, wherever you're from, there's something here for you.

But why should you care? Understanding AI is no longer a luxury; it's a necessity. Just as knowing the internet became crucial, being "AI-literate" is becoming non-negotiable. Understanding AI is about more than just staying updated or fuelling intellectual curiosity. It's about survival and relevance in a rapidly evolving world. You don't need to be a tech whiz, but having a basic grasp can make a difference in your personal and professional life.

So, why this book? Why now? We're not serving a jargon-laden tech primer or a lofty philosophical treatise. Instead, this is a deep dive from first principles, demystifying AI's essence, strengths, and limitations. Through analogies, illustrations, and metaphors, we'll journey through AI's capabilities, understanding its "what," "why," and "how." We'll break down AI's enigma through real-life examples, easy-to-understand stories, and interactive dialogues. As we traverse, expect to gain insights into how AI will revolutionize diverse domains and kindle your creativity and its potential in your sphere of influence.

But with great power comes great responsibility. The impending scale of job loss, the potential misuse of AI, its ethical ramifications, and the dangers of its being weaponized are realities we must confront. This underscores why understanding AI is crucial— harnessing its potential and safeguarding our future. Lastly, to all the parents and educators, preparing our kids for the future intertwined with AI is perhaps the most crucial task.

This book is an effort to ensure that when history looks back at the AI revolution, it sees a society that was informed, prepared, and actively participating. This isn't just a book; it's an experience. An experience that promises to be enlightening, engaging, and empowering. It's an invitation to join a journey to explore, understand, and harness the power and winning with AI.

Disclaimer: In the creation of this book, various AI tools have been utilized to enhance the writing process. While my vision and craftsmanship are the core of the work, the assistance of AI has brought this book to fruition.

Turn the page, and let's dive in!

For further details, deeper insights, or to connect with the author for inquiries, feedback, or collaboration opportunities, simply scan the QR code.

1

UNDERSTANDING INTELLIGENCE

Imagine stumbling upon a magic lamp, only to have a genie emerge, offering to fulfill your every desire. But here's the twist: you can't see this genie. Instead, its presence subtly permeates your smartphone, smart home devices, and online interactions. This invisible force, orchestrating the magic behind screens and algorithms, is called Artificial Intelligence (AI). Much like the mystical genies of folklore, AI captivates our modern world with its seemingly miraculous capabilities. Yet, despite its omnipresence, the true nature of this modern-day marvel eludes many.

AI, once confined to the realms of imaginative tales penned by visionaries like Isaac Asimov, has seamlessly integrated itself into our daily lives. It's as if the fantastical stories of science fiction have leaped from the pages into our living rooms, workplaces, and handheld devices. However, this captivating journey is cloaked in layers of mystery. Utter the phrase "Artificial Intelligence" in a room, and you'll witness a spectrum of reactions ranging from awe and fascination to confusion and skepticism.

Isn't it intriguing? The term "Artificial Intelligence," hailed as the frontier of modern innovation, remains enigmatic to most. It's a well-known play where everyone knows the title, yet only some genuinely comprehend the script. Pose the question "What is Artificial Intelligence?" to your neighbor, a university professor, a tech-savvy friend, or the ambitious founder of a burgeoning AI startup. You'll receive a cacophony of responses, often contradictory.

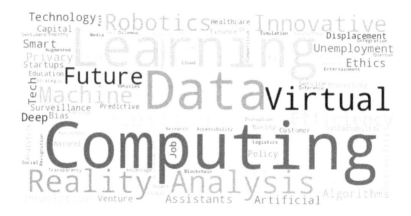

This diversity of understanding is more than just due to the expansive scope of AI but also its chameleon-like adaptability. The term 'AI' seamlessly morphs to fit various narratives, sometimes serving as a mere marketing buzzword while other times embodying a genuine transformative force. Unlike the internet's clear-cut essence of connecting the world, AI resembles a boundless ocean— deep, ever-changing, and challenging to decode explicitly.

Picture assembling a puzzle with infinite pieces, each unique yet contributing to the enigmatic final picture. Now, imagine a puzzle where the ultimate image remains a mystery. This analogy encapsulates the challenge of grappling with the concept of intelligence, especially when appending the term 'artificial.' The complexity multiplies exponentially.

Consider, for a moment, the crystal clarity of an artificial diamond. Its purpose and composition are readily apparent when viewed under the right light. Similarly, when a patient receives an artificial heart, the life-saving function is immediately recognized and celebrated. The clarity stems from the specificity of the original entities they aim to replicate, where purposes are well-understood and defined.

However, intelligence is far from a singular, tangible entity. It's a rich tapestry woven with threads of problem-solving, creativity, emotional understanding, memory, and various cognitive functions. What constitutes intelligence varies from one individual to another, from one culture to the next. While some may view a brilliant mathematician as the epitome of intelligence, others may consider a gifted musician its true representation. Some even argue that a parent skillfully navigating the challenges of raising a child exhibits intelligence at its most profound.

When we endeavor to craft "artificial intelligence," we do not aim to replicate a single, well-defined function. Instead, we're entering a realm where we seek to mimic an array of cognitive processes, many of which we haven't fully understood ourselves. Furthermore, the standards and benchmarks for AI aren't universally accepted. Is a calculator proficient at solving mathematical problems faster than any human considered intelligent? Or does an AI earn that designation only when it can write poetry, discern emotions, or dream?

Herein lies the excitement and challenge of our journey. As we delve deeper into AI, we're not just gaining insights into machines and algorithms. We're embarking on a profound introspection, a quest to grasp the very essence of intelligence. In pursuing artificial intelligence, we inadvertently strive to understand ourselves better. So, while the allure of creating thinking machines may initially captivate us, we're truly engaged in holding a mirror up to the myriad facets of the human mind.

As we gear up to explore Artificial Intelligence in-depth, it's crucial to ground ourselves with a clear understanding of its essence. After all, to navigate a vast ocean, one must first comprehend the nature of its waters. Attempting to explore every application of AI without this foundational understanding would be akin to trying to map every star in the night sky. This noble endeavor may leave us more bewildered than enlightened. Instead, our journey seeks to identify the core, the central pulse of AI, that can serve as our guiding light in delineating the boundaries of this boundless domain.

By pinpointing the heart of AI, we'll chart a clear path forward. This will enable us to distinguish passing trends from transformative breakthroughs and help us cut through the dense fog of hype and marketing bravado.

Philosophical Meanderings and Scientific Enquiries

The concept of intelligence has been a subject of fervent debate and introspection in both philosophical and scientific circles. This exploration traces back to ancient times, with luminaries like Aristotle laying down early foundations. He often associated intelligence with entertaining a thought without necessarily accepting it, emphasizing discernment and contemplation as key aspects of an intelligent mind.

As the centuries unfolded, the understanding and definitions of intelligence underwent significant evolution. The Renaissance, for instance, brought a revival in arts and sciences and a shift in how intelligence was perceived — as a fusion of reason, creativity, and the knack for unraveling the mysteries of the world.

In the modern era, thinkers like Edward Thorndike introduced fresh perspectives. Thorndike proposed intelligence as the 'power of good responses in the context of truth or facts.' Here, intelligence wasn't merely about contemplation or reason but also about making apt decisions and responses when faced with factual scenarios.

Then there was Woodrow, who defined intelligence as 'the capacity to acquire capacity.' His viewpoint, albeit slightly recursive, shed light on the adaptability facet of intelligence. Unlike a vessel with a fixed volume, an intelligent entity could expand its boundaries, adapt, and enhance its capabilities. This revolutionary perspective recognized intelligence as a dynamic attribute that could be nurtured and developed rather than a static quality.

Such viewpoints, among countless others, contribute to a rich tapestry of thoughts on intelligence. They underscore that intelligence is not a monolithic entity but rather a multifaceted gem, reflecting different colors depending on the angle of observation.

As the pages of history turned, humanity's insatiable thirst for understanding the world propelled us into the era of scientific enlightenment. With this shift, the quest to decode intelligence took a decidedly empirical turn. Science, emphasizing measurable evidence and replicable results, sought to quantify that which had previously been primarily qualitative.

Central to this scientific endeavor was the development of the Intelligence Quotient, or IQ score. It represented an attempt to assign a tangible metric to cognitive ability, enabling the measurement of an individual's intelligence relative to the general population. Suddenly, intelligence could be expressed in numbers, ranked, and analyzed, providing an ostensibly objective measure. This had significant academic and socioeconomic implications as IQ scores began to play pivotal roles in educational placements, job recruitments, and beyond.

Yet, as with any scientific inquiry, the field of intelligence assessment continued to evolve. Not all scholars and researchers were content with a single metric attempting to encapsulate the entirety of human cognition.

In a departure from the traditional notion of intelligence as a singular, monolithic entity, Gardner proposed that humans possess various independent intelligences. He identified domains ranging from linguistic to spatial, interpersonal to intrapersonal, and musical to logical-mathematical intelligence. Gardner's theory sheds light on the idea that a mathematician's genius and a musician's virtuosity, while different, are both manifestations of intelligence. This seismic shift suggested that we all possess intelligence in diverse ways, challenging the hegemony of traditional IQ-centric views.

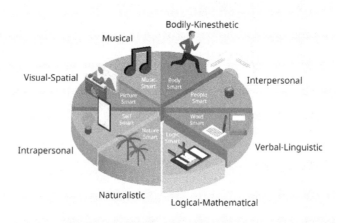

The scientific community's emphasis on quantification extended beyond academic pursuits to practical applications. With quantifiable metrics, educators could tailor learning experiences, employers could optimize workforce allocations, and individuals could gain insights into their strengths and potential growth areas. In a rapidly industrializing world, where efficiency was paramount, understanding intelligence scientifically had ramifications for both individuals and collectives, influencing everything from personal career trajectories to broader labor market dynamics.

Yet, as the chapters of scientific inquiry continue to unfold, it becomes increasingly clear that intelligence remains a vast and intricate puzzle, whether approached philosophically or scientifically.

Distilling The Essence - Toward A Universal Definition

These lenses through which we examine intelligence serve us well in identifying areas for growth, tailoring our learning strategies, and, more recently, guiding us in designing specialized AI algorithms. However, they come with a limitation — the definitions and discussions are intricate and diverse, often pulling us deeper into the labyrinth of the human mind rather than propelling us toward a unified, foundational understanding of intelligence.

A recurring theme that runs through these diverse interpretations is their human-centric nature. Our discourse surrounding intelligence is inherently anthropocentric, founded upon the premise that this remarkable trait belongs exclusively to Homo sapiens. However, as we step into a new era - one in which we strive to comprehend and construct intelligence beyond the confines of the human brain - we are compelled to broaden our perspective.

Imagine a prism. When white light passes through it, it splinters into a spectrum of colors, each distinct and representing a unique light facet. In our journey, we have extensively explored each hue and manifestation of intelligence. But now, our quest has taken a different turn. We aspire to synthesize these manifold colors into a singular beam of white light, a comprehensive representation of intelligence.

The question arises: Can intelligence be distilled into a foundational concept? Can it be universally defined without being tethered to human distinctive characteristics?

To grasp the essence of intelligence, we must trace our steps back to the inception of life itself. It's a voyage that leads us from the primordial soup of existence through the intricate dance of organic compounds to the elegant ballet of cellular life and beyond to the myriad of organisms that grace our planet today.

Indeed, science has granted us the privilege to peer into the most intimate realms of life. We have scrutinized the mesmerizing choreography of molecules and cells under the unblinking gaze of our microscopes. We have dissected life to its basic unit, the cell - a marvel of Nature's engineering, a symphony of atoms and molecules orchestrated into a vibrant, pulsating entity. Yet, the profound questions that stir our souls remain. How did this all come to be? How did this cosmic ballet of atoms and molecules culminate into cells? How did these cells intertwine to spawn the incredible complexity and diversity of life as we know it?

To answer these questions, let us envisage an invisible hand guiding the evolution of life - a phenomenon we can term the 'intelligence of life.' This is not intelligence in the traditional sense but rather a fundamental property. This primal force propels life to evolve and improve incessantly.

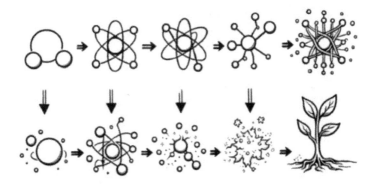

Imagine a simple rule of the universe: 'Create something better using the elements at your disposal.' This basic yet profound principle delineates the living from the inert. This principle, this intelligence of life that catalyzes the amalgamation of molecules into cells, orchestrates cells into complex organisms, and choreographs the intricate dance of evolution, enabling life to diversify, specialize, adapt, and thrive.

This invisible force elucidates the grandeur of life's complexity - the symphony that ensues when multiple entities interact, yielding a whole that surpasses the sum of its parts. It elucidates how the rudimentary, single-celled amoeba, given ample time, could evolve into the myriad forms of life we witness today, from the humble worm burrowing in the soil to the majestic blue whale traversing the vast oceans, and even to us, Homo sapiens, capable of pondering our origins.

Contrast this with the realm of inert. A rock, despite enduring for eons, remains just a rock. It neither grows, evolves, nor adapts. It harbors no aspirations to transcend its current state. This is the distinction that the 'intelligence of life' bestows upon the living. It infuses purpose into atoms and molecules, directing them to cooperate, innovate, and evolve. It differentiates a mound of inanimate matter from a living, breathing organism. Let's perceive intelligence not as a unique human trait but as a universal principle.

This propulsive force propels life's grand odyssey from the tiniest microorganism to the most advanced human beings.

As we delve deeper into this exploration, a universal rhythm pattern emerges—a natural progression characterized by 'Doing more with what we have.' This rhythmic cadence is the underlying beat orchestrating the grand symphony of evolution and the diverse spectrum of human technological innovation. It embodies the phenomenon of compounding incremental developments, a snowball effect that, over time, culminates in significant leaps forward.

But what fuels the flames of innovation, evolution, and intelligence? What propels this ceaseless march toward improvement? The answer lies in a fundamental driving force: 'Incentives.' Incentives are the lifeblood that infuses vitality into the body of life's activities, orchestrating the intricate ballet of biological and technological evolution. The quest for survival, the desire to thrive, the aspiration to perpetuate across time—these fundamental incentives ignite the intelligence engine. Whether organisms adapt to their environment, humans invent vaccines or computers, or amoebas seek sustenance, the driving force behind these actions is the incentive. This principle holds for the humblest microorganisms and the most advanced life forms—humans included.

Intelligence can be understood as an entity's response to its incentives, a flexible and adaptable tool shaped by the environment and the challenges it presents. The more sophisticated this response is, the higher the level of intelligence is and, consequently, the greater the domination of life on this planet. In the early days of humanity, intelligence manifested as the capacity to develop agriculture and invent tools like the plow. A few decades ago, being able to perform swift mental calculations or possessing a vast vocabulary was deemed a sign of high intelligence. However, we now understand that these definitions of intelligence are not absolute but reflections of the environment's incentives over time.

Thus, intelligence emerges as the essence of life, the secret ingredient that sets the living apart from the inert, the driving force propelling life forward in its relentless pursuit of improvement. This perspective allows us to interpret all facets of intelligence—problem-solving, learning capacity, adaptability—not as distinct phenomena but as different expressions and layers of the same fundamental drive to survive, thrive, and perpetuate. In this context, it becomes evident that machines do not fit into this facet of intelligence, as they lack the innate desire or incentive to thrive and perpetuate.

Artificial Intelligence - A Misnomer

As we venture further into intelligence, a fundamental revelation emerges: Intelligence is relative, rooted in context and time. It would be erroneous to assert that we, in our age of sleek smartphones and quantum computers, are inherently more intelligent than our grandparents, who navigated life without the assistance of these digital companions. We are equipped with different tools tailored to our present circumstances, and the skills required to wield these tools are distinct from those needed in previous eras.

This contextual nature of intelligence poses a significant challenge when applied to non-living entities, giving rise to what many consider a 'misnomer'—artificial intelligence. The term evokes images of machines mirroring human intelligence, echoing our thoughts, and emulating our actions. However, the reality is somewhat different.

Machines operate differently and lack life's primal drivers—the incentives to survive, thrive, and perpetuate. They serve as tools in humanity's hands, enhancing our capabilities, expanding our reach, and unlocking our potential.

While they have undoubtedly grown more sophisticated, this sophistication does not equate to the same form of intelligence found in living entities.

It would be more apt to refer to this phenomenon as 'Machine Intelligence' rather than 'Artificial Intelligence.' A subtle shift in terminology can help us dispel the illusions surrounding this field and cultivate a clearer, more nuanced understanding.

By advancing machines' capabilities to respond dynamically in a non-preprogrammed manner within their boundaries of knowledge, we're witnessing a monumental leap in technological prowess. While this certainly warrants the use of the term "intelligence," the term "Artificial" implies a man-made entity that will mimic the innate abilities of "living" entities, human beings included.

Machine Intelligence - Magic of Dynamic Responses

If you've ever observed a grandfather clock, you would have noticed the steady tick-tock of the pendulum. Each movement is predetermined as a result of meticulous engineering. For centuries, every piece of machinery, from towering cranes to complex computers on our desks, operated in this manner—slaves to their predefined blueprints. In their earliest iterations, traditional machines were mere extensions of human physical prowess. Consider vehicles: intricately designed yet reliant on human inputs, with every turn, acceleration, or brake dependent on the driver's command. Similarly, cranes, ships, and pumps represented sophisticated tools, magnifying our physical capabilities yet bound by human direction.

The invention of computers marked a paradigm shift. Instead of merely aiding physical tasks, they began to complement our cognitive functions. With their ability to crunch numbers, archive texts, or even stream media, these digital wonders were impressive but still limited. They were designed to operate within the confines of human-set programs, simply executing tasks they were instructed to do.

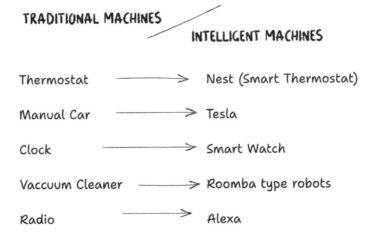

TRADITIONAL MACHINES

INTELLIGENT MACHINES

Thermostat ——————→ Nest (Smart Thermostat)

Manual Car ——————→ Tesla

Clock ——————→ Smart Watch

Vaccuum Cleaner ——————→ Roomba type robots

Radio ——————→ Alexa

The concept of "Intelligent Machines" emerges as a revolutionary leap from this backdrop. Pre-set, static operations do not shackle such machines. They demonstrate an extraordinary ability to adapt, responding to conditions beyond rigid programming. For instance, while a traditional car maneuvers solely based on its driver's inputs, a car equipped with "automatic cruise control" adjusts its speed based on the behavior of vehicles around it, embodying the essence of "machine intelligence."

But what sets apart a mere machine from an intelligent one? It's not just about performing a task but adapting to unforeseen circumstances. It's about responding dynamically to inputs that haven't been explicitly programmed.

Consider a thermostat. In its conventional form, you set a temperature, and it heats or cools the room accordingly. Now, introduce a thermostat with Machine Intelligence. Not only does it adjust the room's temperature based on your setting, but it also learns your preferences over time, anticipates when you'll be home, checks the outside weather, and even senses when the room is occupied. This thermostat doesn't merely follow orders; it interprets, predicts, and decides.

Drawing upon this, it's evident that the essence of Machine Intelligence lies in its adaptability. While traditional machines are akin to actors rehearsing a well-practiced script, intelligent machines are like improv artists, adept at responding to the unpredictable.

Any software or hardware embedded with systems enabling dynamic responses with non-pre-programmed behavior can be considered within the "Machine Intelligence."

At first glance, machines have been displaying intelligence for decades. So, what's all the contemporary excitement about? What makes these smart thermostat-like systems so intriguing? Just as in the realm of living organisms, where there's a vast difference in intelligence levels between an amoeba and a human, the world of machines spans a wide spectrum.

Consider earlier machine advancements: lane assistance in cars, automatic intrusion detection in wifi routers, or even the versatile robot vacuum cleaners that navigate any room layout. While these feats were undoubtedly impressive, they were somewhat limited in complexity. Each system responded dynamically, yes, but within a specific, often narrow domain. It's akin to the intelligence exhibited by lower organisms – commendable, yet not groundbreaking.

The enthusiasm surrounding the current era of machine intelligence isn't merely due to machines responding dynamically. It's about the sheer breadth and depth of their responsiveness. It's about machines that don't just perceive our words but grasp their underlying nuances. It's about systems that don't just recognize faces but can interpret emotions conveyed through subtle facial twitches.

What we're witnessing now is a profound leap. Contemporary machine intelligence has broadened its scope and achieved a nuanced understanding of the wide diversity and complexity of aspects traditionally considered human strengths. We're discussing language, music, videos, software, and more. Having been trained on a vast corpus of data, these systems can find the best combinations to respond to new situations.

Imagine an extraterrestrial observer charting the course of evolution on Earth. The emergence of an amoeba, while a significant leap from non-living matter, might only pique mild interest. But the dawn of Homo sapiens? That would undoubtedly be a moment of intense intrigue, marking the threshold of endless possibilities. It was not just about the birth of "Cavemen" but their potential to become "Space explorers." This analogy aptly encapsulates today's machine intelligence.

However sophisticated they become, they will still require the primal intent and drive to act autonomously. They will be very smart tools assisting humans. In the future, we could see some smart machines whose actions could be dynamic and respond to unseen input at an unprecedented scale. However, their intent will still be programmed and controlled by humans.

Our journey through the enthralling world of 'intelligence' brings us to a juncture where we understand the need to redefine our terminologies to match our understanding. Yet, due to its charm and allure, we recognize the widespread acceptance and use of the term' Artificial Intelligence.' It symbolizes "marketing success," which immediately draws our attention. Despite our exploration revealing it as somewhat of a misnomer, its widespread usage in academic, industry, and popular terminology cannot be overlooked.

With the notion of 'Machine Intelligence,' we have now drawn a more accurate picture, distancing ourselves from the lofty imaginations fostered by Hollywood and sensationalist media. It provides a platform to keep our feet on the ground while our minds explore this technology's immense possibilities.

However, for continuity and coherence with the world outside this book, we will continue to use the term' Artificial Intelligence' or

'AI.' This foundational understanding will guide us as we journey further into the myriad dimensions of AI. We embark on this adventure with a fresh perspective, equipped with the clarity that the term 'AI' doesn't conjure beings of consciousness sprung from lines of code but incredible tools with which humanity can shape its future.

2

UNDERSTANDING AI

Consider this: You are tasked to design a house, but instead of drafting the blueprint yourself, you lean over to your computer and specify:

Create a dwelling of dreams, a two-story masterpiece where modernity intersects with nature. An abode that whispers the melodies of the coastal winds and embraces the warmth of tropical sun yet stands steadfast against the seasonal rain. Tailor it for those who relish opulence yet seek a humble harmony with the Earth.

You sit back, sip your coffee, and watch in awe as your computer sketches an architectural masterpiece, assembling various elements like foundation, doors, walls, windows, and roof in a logical and aesthetically pleasing manner.

Sound too good to be true?

Welcome to the fascinating world of AI! Though this particular illustration might not exist when writing this book, it's only a matter of time before it becomes a reality.

To understand how this is possible, consider each part of a house - the foundation, basement, entrance, doors, windows, rooms, roof, balcony, courtyard, garage, and walls - as "Components." These components are the building blocks, the Lego pieces, if you will, that the computer uses to construct a house.

Imagine we have an extensive catalog of thousands of house models designed by top architects worldwide. This vast catalog is called "Data" in the AI world. Each model is marked and labeled meticulously, with each component in its place. We hand this vast catalog to the computer, which processes these models to understand the spatial relationships and configurations between the different components.

Like an eager student studying for a major exam, the computer determines the probabilities of these relationships. It learns that the foundation is always at the lowest level, the basement comes next, floors are horizontal, doors are typically vertical at the intersection between floor and walls, and so forth. This learning process gives the computer a 'sense' of how to construct a house logically.

Here's where the magic happens: once the computer has figured out the probabilities of these configurations, we can ask it to generate a house with specific parameters—for instance, a two-story house with a grand main entrance.

The computer starts with the highly probable components – the foundation on the ground, the doors intersecting floors and walls – and then brings variations based on lower probability components. For instance, it might decide whether the stairs should be inside or outside or if the roof should be flat or sloping.

Much like a master painter who mixes a finite set of colors to create an infinite variety of paintings, our computer, given the diverse components, generates endless permutations and combinations of houses, all in a manner that 'makes sense.'

The secret to its success? Studying and learning from the vast catalog of human-designed houses.

But wait, there's more! Let's sprinkle some extra "contextual information" or "Factors" into the mix. Suppose we tell the computer that houses with sloping roofs are typically found in regions with heavy rain or snowfall. At the same time, those with cross-ventilation are suitable for hot climates.

We might also indicate that larger houses with more rooms are generally associated with higher budgets. In contrast, smaller ones correspond to lower budgets. These factors give the computer additional rules or guidelines when designing houses. So, the next time you ask it to create a house suitable for a tropical climate with lots of rainfall, it knows exactly what to do for a wealthy client.

Welcome to the future of design powered by AI. By combining components in a manner that makes sense and adjusting them based on contextual factors, this technology gives us the power to generate an endless array of designs, opening up a world of architectural possibilities limited only by our imagination. Taking our metaphor further, the latest advancements in AI resemble a supercharged computerized architect. This cutting-edge technology isn't limited to learning the probabilities of relations between the components of a few house models. Instead, it's capable of mastering a vast number of components across countless factors.

The future of AI promises a world where computers can help us create designs beyond our wildest imagination while adapting to specific, unique requirements. It's like having a master architect at our fingertips, available around the clock to bring our visions to life.

How exciting is that? While our metaphor of designing houses with AI remains in the realm of imagination, similar feats are already a reality in text and image generation domains. AI can write coherent paragraphs, create original art, and even compose music. This is no longer just a figment of science fiction; it's transforming our reality.

Simple Essence of AI - The Interplay of Data and Labels

For all the magic that AI can do, at its heart, it is a sophisticated game of connecting the dots. AI is a complex probability model that identifies the relationships between data and its factors. If one were to distill the essence of AI, it would be this interplay between the data, often vast and varied, and its corresponding labels. Understanding this core mechanism illuminates the vast expanse of possibilities that AI promises.

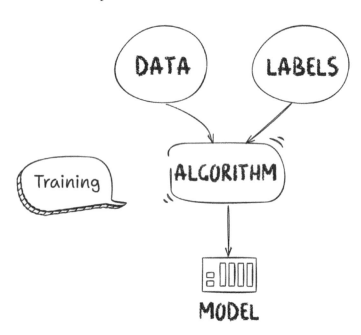

Let's delve deeper into this. When we talk about AI in the context of "Graphics," the data predominantly consists of many images. These images, in turn, are associated with labels. Think of these labels as tags or descriptors we assign to different pieces of data. These labels are descriptive attributes highlighting the images' properties, utilities, and characteristics. For instance, an image could be labeled as "cat," "van Gogh style," "sky," "modern city," or "photo-realistic." Essentially, any attribute a human mind can associate with an image can be its label.

Similarly, billions of lines of text form the data in textual AI. Each line or paragraph has its respective labels that provide context, meaning, or categorization. These labels are pivotal as they help the AI system understand and interpret the content.

The beauty of AI is its simplicity. Once you've built this intricate map of data and labels, you can perform only two kinds of operations. The first operation involves presenting the AI with new data and querying it to identify and return labels. When applied to data such as images, text, or numeric values, this operation is termed "Classification." However, if the labels are time-based, this process is aptly called "Prediction." Collectively, this realm of AI is termed "Discriminative AI". Imagine feeding a picture of a skyline into the system. Discriminative AI would identify and label elements like "skyscrapers," "sunset," or "crowded." It discriminates or classifies based on patterns it has learned from previous data. On the other hand, when the labels relate to time, this operation transforms into prediction. For instance, based on historical stock prices (the data), an AI might predict future prices (the labels).

CLASSIFICATION

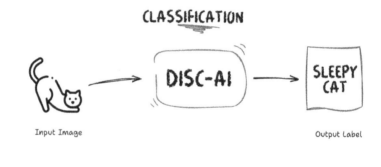

Input Image   Output Label

PREDICTION

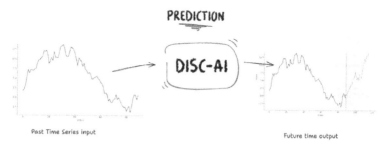

Past Time Series input   Future time output

Conversely, the second operation is more creative. Here, one provides the AI system with a set of labels and tasks it with crafting a new piece of data that aligns with those labels. Here's where the tables turn. Instead of giving AI data, we provide it with labels or instructions. We might tell it, "Paint me a digital landscape with a 'Van Gogh style,' 'cat,' and 'sky.'" AI then crafts a brand-new piece of data based on those labels, an image, or a line of text. This act is about creation and imagination. This fascinating domain is known as "Generative AI."

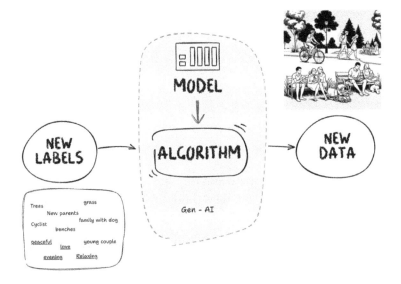

The current excitement in the AI world is largely due to the sheer diversity in the volume of data and the diverse range of labels we can now handle. A comprehensive understanding of the magic of AI will come from something other than the complex maze of jargon. Still, through the lens of this straightforward relationship, AI is fundamentally a probabilistic model intertwining data with its labels. And it's this simplicity, applied to vast datasets and a plethora of labels, that has opened the door to countless applications that are only limited by human imagination. To absorb and appreciate this ingenuity, let's traverse the short history of AI, where it started, and how it has evolved to the present day.

Evolution of AI

The innate human qualities of curiosity and exploration, along with a relentless drive to streamline tasks and improve efficiency, have consistently fueled the evolution of technology throughout history. The advent of computers sparked a transformative wave, opening up a vast landscape of unprecedented potential. Given that we inhabit a primarily visual world, one of our most fundamental abilities lies in recognizing patterns, an instinct deeply ingrained and indispensable. We, therefore, embarked on a journey to harness the power of computers in this domain, aiming to mimic our inherent capacity for pattern recognition. This marked the genesis of the practical narrative of Artificial Intelligence as we perceive it today. A new chapter in our technological odyssey is characterized by the intricate replication of human cognitive faculties within the digital framework of machines.

Gazing upon a computer and seeing a mind was the beginning of AI.

One of the first and most profound applications was in calculations. Computers could execute arithmetic operations with a speed and accuracy far surpassing human capabilities. This revolution in computation sparked a revolution in thought: If machines could outperform us in such a quintessentially human task, what else could they do?

This breakthrough ignited the innate curiosity of humankind. We began to wonder and imagine the vast range of mental capabilities we could imbue into our machines. What if a machine could calculate, learn, reason, and make decisions? What if a machine could understand and respond to human language? What if a machine could simulate human intelligence? The questions were as endless as they were intriguing, propelling us onto the path of artificial intelligence. A path that, as we shall see, has proven to be as transformative as it is challenging.

The world we inhabit is a constant whirlwind of information, swirling around us in vibrant hues and complex patterns. Our very existence hinges on our ability to perceive and interpret these myriad visual cues embedded in our environment. Vision, therefore, is much more than just a sensory function. It's our window to the world, a cornerstone of our consciousness, and plays a crucial role in our survival and prosperity.

When we delve deeper into the mechanics of vision, we realize that it involves a sophisticated process of deconstructing the visual inputs received through our eyes into a set of semantic labels. Our brains then weave these labels to form coherent and meaningful interpretations of our surroundings. We see colors and shapes, objects, entities, and their relationships. We detect motion and anticipate trajectory, understanding the world in space and space-time.

This ability to recognize and make sense of patterns is an innate human capability deeply ingrained into our DNA, a testament to millions of years of evolutionary adaptation.

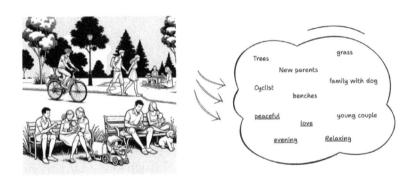

With the advent of computers, this primal human capability sparked a new avenue of exploration. If humans, with our biological limitations, could master the art of pattern recognition, why couldn't our creations? Thus, we embarked on a quest to make machines comprehend the visual world, much like we do.

Our initial endeavors were humble. The story begins in the late 20th century with the U.S. Postal Service (USPS) employing simple optical character recognition (OCR) algorithms. These algorithms, capable of interpreting handwritten or typed text on envelopes, streamlined the process of sorting mail, significantly improving the efficiency of the postal system. It was a modest beginning but an essential step towards more complex visual processing tasks.

Next, we turned to the industrial sector. Automated inspection systems started to use machine vision to identify manufacturing defects. This improved the quality of the products, saved a significant amount of time, and reduced human error. In the automotive industry, OCR technology evolved to read and interpret number plates, paving the way for automated parking tickets and toll collection systems and improving road safety through surveillance.

Meanwhile, in medicine, machine vision began transforming diagnostic procedures. Simple X-ray imaging evolved into machines capable of identifying fractures, abnormalities, and diseases. Eventually, complex imaging techniques like MRI and CT scans were paired with sophisticated machine learning algorithms to detect subtle anomalies, such as tumors, that the human eye could easily overlook.

Security and surveillance systems also saw significant improvements. Facial recognition technology began to take shape, initially identifying individuals in black-and-white images, later advancing to full-color recognition in dynamic, real-time environments. These systems were then used to enhance security measures in various public spaces and institutions, from airports and banks to corporate offices. Let's see how the technology behind these applications evolved.

The journey of machine vision, from the simple task of recognizing black-and-white characters to the intricate analysis of complex scenes, intriguingly parallels the development of visual perception in humans. Just as a newborn gradually learns to make sense of the world around them, machine vision also matures and evolves, deepening its understanding of the visual world. A human infant begins life with a fuzzy, indistinct view of their surroundings. Over time, they learn to differentiate between black and white, high-contrast colors, and movements. As their cognitive abilities develop, they recognize basic shapes and a spectrum of colors. Gradually, they begin to perceive and understand more complex scenes. The development of these skills in babies and machines share striking similarities, suggesting that machine vision follows the trajectory of human visual cognition.

In its early stages, machine vision could identify rudimentary features like edges and corners, much like a young child recognizing simple shapes. This collection of identifiable features was then classified into patterns, a foundational step toward more advanced recognition tasks.

As computational power increased, the complexity of features that machines could recognize also grew. This growth mirrors how a child's visual cognition expands as their brain develops. Machines were manually trained to identify complex features such as facial characteristics, using rules like the presence of a pair of eyes, a nose, a mouth, and ear-like structures within specific dimensions to classify an image as a human face.

However, this approach hit a roadblock like a toddler needing help comprehending a complex scene. The handcrafted feature identification method showed limitations in reliability, especially when handling intricate scenes. It became increasingly clear that creating value was becoming difficult against the backdrop of rising computational complexity.

This bottleneck was akin to a child reaching a point where their current level of understanding isn't sufficient to interpret the world around them accurately. It represented a turning point in the evolution of machine vision, where a shift in approach was needed to break through the limitations and move towards more sophisticated capabilities.

With the limitations of handcrafted features and a growing desire to push machine vision to new heights, scientists started looking towards a more profound source of inspiration - the human brain.

Their quest was to replicate human capabilities and understand how humans process a vast array of complex and dynamic scenes with ease and proficiency. They found the key in the intricate networks of nerve cells, or neurons, that make up the brain's fabric.

The human brain is home to a complex and sophisticated network of neurons, an intricate system that evolved over millions of years. A baby, starting with a neural network like a blank canvas, embarks on an incredible journey of learning and development. This neural network grows increasingly intricate, with new neurons formed, existing ones connected in novel ways, and some old connections pruned. This neural evolution, which begins in infancy and continues throughout life, allows us to easily understand and interpret increasingly complex scenes. This remarkable evolution is fueled and perpetuated by our brains being continuously bombarded with new information.

Intrigued by this process, scientists started working on replicating this organic, adaptive learning mechanism. The result was the birth of artificial neural networks, computational models designed to mimic the human brain's structure and function. They started by creating simple networks, gradually adding more layers and connections to increase complexity, much like the growth of a child's neural network.

However, just like a child's brain needs a steady stream of new experiences to learn and adapt, these artificial neural networks must also be fed information to evolve. This information, in the form of data, was their learning material, their lifeblood. Without it, even the most sophisticated neural network would remain a blank slate, unable to discern even the simplest patterns.

Yet, while researchers had succeeded in creating these artificial neural networks, they found their utility constrained by data availability. Without a sufficiently large and diverse dataset to train on, these networks could not truly mimic the sophisticated visual recognition abilities of the human brain. In the initial stages, the utility of artificial neural networks was limited by data availability. After all, a network that isn't exposed to varied information can't learn effectively, just like a child in isolation wouldn't develop robust cognitive abilities. This powerful idea hibernated at that time due to the limitations of computational capability and data availability.

Over the past two decades, several independent advancements converged to facilitate a revolution in artificial learning. The ubiquity of digital imaging technologies and social networks, coupled with the migration to cloud storage, brought about an explosion in data creation and accessibility. This profusion of data reinvigorated interest in artificial neural networks – systems capable of autonomously adapting based on the data they were exposed to without needing specific handcrafted features.

An important milestone in this journey was a neural network's successful identification of cats in various contexts. Though this might be a funny story, this achievement proved that these data-driven systems could effectively recognize complex patterns and understand intricate scenes. Propelled by this breakthrough, the race was on to design more advanced and powerful networks.

The increase in computational power, brought about by advancements in graphics processing units (GPUs) and the spread of cloud computing, acted as a tailwind in this race. Networks grew increasingly complex, extending to many layers to understand the massive diversity of data. Thus, these advanced massive artificial neural networks became known as "Deep Neural Networks." Their learning mechanisms, characterized by their ability to extract intricate patterns from massive datasets, were aptly termed "Deep Learning."

The Feline Revelation

The year 2012 marked a significant leap in this field. The global AI community held its breath when Google's secretive X Labs announced they had built a network of 16,000 computer processors and used it to train an artificial neural network on 10 million YouTube clips.

The task was ambitious: to see if this artificial brain could identify patterns and recognize objects from the colossal amount of unlabelled data it was presented with.

After three days of relentless processing, the artificial brain had a breakthrough – it learned to recognize a cat. While seemingly ordinary, this feat marked a monumental stride forward in artificial intelligence.

But why was this discovery so significant? After all, it's no small feat to engage 16,000 processors just to recognize a household pet. However, the implications of this experiment stretch far beyond the realm of 'feline' recognition.

Firstly, identifying cats in unlabelled YouTube clips proved that artificial neural networks could learn independently. The network was not preprogrammed to look for cats or any specific object. Yet, it discovered and recognized them solely based on the frequency of their appearance. This ability to learn from unstructured data and draw conclusions without explicit guidance closely mirrors how human brains learn, making this a momentous leap toward creating machines that can truly mimic human cognition.

Secondly, the scale of the operation was an important demonstration of the sheer computational power required to process such a vast amount of information. It confirmed many AI researchers' hypothesis: the larger and more powerful these models, the better their performance.

While employing 16,000 processors to recognize cats may seem excessive, the experiment was not about cats per se. As Andrew Ng, one of the lead researchers and pioneers of the field, noted: It was about testing the capabilities of a vast, intricate neural network – a 'simulated baby brain' – to learn from observing the world, just like a human child would.

The cat experiment clearly vindicated this approach. Despite its massive data consumption and computational requirements, the fact that the artificial neural network learned to recognize something as complex as a cat without any specific guidance was nothing short of extraordinary.

In essence, Google's experiment underscored a crucial truth: the road to developing truly intelligent machines requires a blend of sophisticated models, colossal data sets, and an unprecedented level of computational power. The cat experiment was just the beginning – the first step in a long journey towards creating machines that can learn, adapt, and comprehend the world around them as seamlessly as humans do.

Within a few years, complex neural networks would start recognizing several thousand categories of objects within a given image, mirroring the learning process in human babies. While we talk about the evolution of "Image understanding technologies," a similar trajectory echoed in the field of "natural language understanding." With more than 1.86 billion websites worldwide in 2021, the amount of data available has grown exponentially, assisting the growth of AI systems that can understand patterns in natural language.

Discriminative AI - The Art of Classification

Since that pivotal moment, the capabilities of these AI systems have grown exponentially. They can now recognize and understand an ever-increasing array of objects and concepts. This advancement gave birth to what we now call "Discriminative AI." These AI systems excel at distinguishing different categories based on the input data. They can dissect a scene, identify its constituent elements, and categorize them, making sense of the visual world.

It seeks to identify or classify the given input – an image, a block of text, an audio snippet, or a more complex scene – into distinct categories. The term "discriminative" stems from the AI's ability to discriminate or differentiate between various inputs based on their characteristics. For instance, a discriminative AI model might be given a photograph and tasked with identifying whether it's an image of a cat or a dog. It would analyze the patterns present in the input – the shape of the ears, the texture of the fur, the color of the eyes – and classify the image accordingly. This is an example of AI mimicking the human ability to recognize and categorize objects in our environment, a fundamental aspect of how we interact with the world around us.

It analyses the characteristics of the given input. It makes informed decisions about what category it falls into, like a human brain identifying and categorizing objects, videos, or sounds based on their features. This principle of Discriminative AI has broad applications across several domains, including image processing, video classification, and natural language understanding.

In the realm of image processing, Discriminative AI models excel at distinguishing between various categories. Given a dataset of images, a discriminative model can learn to identify the presence of specific elements, such as animals and buildings, or even more abstract concepts, like emotions conveyed by facial expressions. The key lies in training the model with a vast range of labeled images. As it sifts through this wealth of data, the AI learns to discern patterns and characteristics unique to each category. For instance, if trained with images of cats and dogs, the AI learns to discriminate between the shape of the ears, the color of the fur, or the animal's size, helping it accurately categorize new, unseen images.

In video classification, the application of Discriminative AI becomes even more dynamic. Instead of a static image, the AI now deals with a sequence of frames, each adding to the complexity of the data. In addition to recognizing objects in each frame, the AI must also understand their movement and interaction within the video. For example, a discriminative model trained in sports analysis might need to distinguish between different types of sports. This might be a difference between "baseball" or "tennis". It doesn't just identify the presence of players and equipment. Still, it recognizes the unique patterns of movement, the rules, and the game's flow to categorize the video correctly. The technology has advanced to the scale of connecting these individual tracks, for example, "In tennis, tracking Arms, legs, racket and the ball," over many video streams to get detailed statistics about a player's gameplay and strategy.

NLU (Natural Language Understanding) is a subfield of AI that deals with the interaction between computers and human language. It's about enabling machines to understand and respond to inputs in natural language forms, similar to humans. The role of Discriminative AI in NLU is of fundamental importance and is akin to its role in machine vision – it classifies and categorizes text or spoken language into predefined classes to understand and make sense of them. This can range from sentiment analysis (classifying text as positive, negative, or neutral) to more complex tasks like understanding the intent behind a user's query in a conversational AI system and effective translations. For example, consider a machine tasked with sentiment analysis, which is determining whether a piece of text is positive, negative, or neutral in sentiment. Given a statement like "The service at the restaurant was fantastic," the AI must classify this statement as positive. It discriminates the sentiment of the input by recognizing specific patterns, not just the presence of the word "fantastic" but understanding its connotative meaning within the context of the sentence.

While images offer visual data ripe for pattern recognition, text presents a different challenge for AI systems. When faced with a body of text, a Discriminative AI model doesn't just see a series of letters and punctuation marks; it sees an intricate fabric of meanings and relationships, just as humans do when reading. To accomplish this, these models must be trained to recognize and interpret the subtle patterns in language – the syntax, the semantics, and the intricate web of relationships between words, sentences, and larger bodies of text. They need to comprehend the roles words play in a sentence, discern the meanings of idiomatic expressions, detect the subtle shifts in tone, and understand references and contextual cues.

For example, consider a model trained to classify customer reviews as positive, negative, or neutral. The Discriminative AI would analyze the text of each review, identifying key phrases, sentiment-indicative words, and overall tone to categorize each review accurately.

A phrase like "highly recommend" would push the review towards the "positive" category. At the same time, "terrible service" would swing it toward the "negative" side. Moreover, Discriminative AI models have found crucial applications in complex tasks like topic modeling, where the AI identifies the main themes present in a text, Named Entity Recognition (NER), part-of-speech tagging, and semantic role labeling. In NER, the model identifies names of people, organizations, locations, and other specifics in a given text. In part-of-speech tagging, the AI classifies words into their respective parts of speech – noun, verb, adjective, etc. On the other hand, semantic role labeling is about understanding the relationships between entities and actions in a sentence.

In more complex applications like language translation, the AI must understand the meaning of each word in the context of a sentence in the source language, then classify and translate it into the equivalent words and sentence structure in the target language. This process involves recognizing patterns, classifying inputs, and making sense of both languages' syntax, semantics, and pragmatics. Take the sentence, "John tossed the ball to Mary in the park." An AI model performing semantic role labeling would identify "John" as the one acting (tossing), "the ball" as the object being acted upon, "Mary" as the recipient of the action, and "the park" as the location where the action took place.

Discriminative AI is the Sherlock Holmes of the digital world; it investigates, classifies, and discerns with a cold, calculated logic that would make even the great detective tip his hat.

Discriminative AI's role in NLU underpins many AI-powered systems with which we interact daily. From voice assistants like Siri or Alexa, which understand our voice commands, to email filters, which categorize incoming messages as "spam" or "non-spam," Discriminative AI is constantly at work. By leveraging patterns and applying them in different contexts, Discriminative AI enables us to build systems that understand, respond to, and even predict human language in nuanced ways, transforming how we interact with technology. Across these domains, the principle of Discriminative AI remains the same: given an input, discriminate it into one of several possible categories. This ability to accurately categorize and classify has broad implications for AI's role in diverse fields, from healthcare to entertainment, making Discriminative AI a crucial pillar of the AI landscape.

But the story of artificial intelligence isn't a one-way street. Just as the human intellect refuses to stagnate, the development of AI continues to innovate, constantly exploring new territories. If an AI system could take an image of a cat and generate the label "cat," what would happen if the input and output were swapped? This persistent curiosity led to a fundamental question: Could this process be inverted? Could we prompt an AI system with a description and expect it to produce a corresponding image?

Generative AI - Crafting the New from The Known

This thought experiment gave birth to "Generative AI," a new frontier in artificial intelligence. Unlike discriminative AI, which categorizes or classifies inputs, generative AI creates or generates outputs based on a given input. Imagine you input the phrase, "Generate an image of a cat in a bathtub," a machine can create an entirely new image representing the requested scene. This AI isn't simply categorizing or recognizing existing patterns; it's generating new content, mirroring the human ability to imagine and create.

While Discriminative AI seeks to understand and categorize the world, Generative AI strives to create new, novel content from a given input. The input and output can span multiple modalities— images, audio, video, or text—expanding the potential applications of AI in fascinating ways.

Generative AI models are used to create meaningful and coherent text sequences in the context of language. This can be as simple as generating a sentence or as complex as writing an entire novel. The goal is for the AI to produce text that is not only grammatically correct but also contextually relevant and creative, akin to human writing.

For instance, consider the case of a machine learning model tasked with writing an article. Given a title or a brief prompt, the Generative AI uses its knowledge and understanding of language, learned from extensive training on a vast dataset, to generate a coherent, contextually relevant, and informative article. It's not merely regurgitating learned phrases but combining words, sentences, and paragraphs to create new content.

A Generative AI model could be given the prompt, "Write a short story about a boy who discovers a magical land." The model would then generate a new, original story that follows the given premise. It's essentially crafting a narrative, weaving characters, and describing scenes – all the components of creative storytelling – purely based on a simple text prompt.Gen-AI is the unseen artist composing masterpieces from the mundane like the symphony born from silence or the poem birthed from a blank page.

Another application of Generative AI in the realm of NLU is chatbots. While early chatbots could only respond to specific prompts with preprogrammed responses, modern AI-powered chatbots can generate unique responses to various queries. They understand the context of the conversation, remember past interactions, and create responses that align with the ongoing dialogue – all thanks to the generative capabilities of AI.

Generative AI models learn from vast amounts of data in all these applications, capturing human language's nuances, style, and structure. They then use this knowledge to generate new, unseen content.

They're like diligent students, learning from millions of books, articles, and conversations and using this learning to write their narratives. Generative AI threads the creation narrative in the fabric of creation and perception, enabling machines to mirror the human capability to invent and imagine. By bridging the gap between understanding and creation, Generative AI is shaping a future where machines are not just tools for analysis but also partners in creation.

The story of Generative AI does not end here. It has just begun. This has sparked an exciting new era for AI and is at the root of widespread interest and public discourse. In future chapters, we will dive deep into understanding this groundbreaking technology and its vast implications.

GAN - The Convergence of Creativity and Critique

Imagine a crime mystery novel featuring two main characters: a masterful counterfeiter and an astute detective. The counterfeiter's job is to create perfect counterfeit money that looks identical to real currency, and the detective's mission is to distinguish real money from fake money.

The counterfeiter starts his operations and passes the forged money around town. Using his keen eye and expertise, the detective looks at each piece of currency, determining if it is real or counterfeit. Whenever the detective finds a counterfeit bill, he studies it, identifying the telltale signs that gave it away. He then provides this feedback to the counterfeiter, not intending to arrest him but to push him to improve his forgery skills.

Encouraged by the challenge, the counterfeiter uses this feedback to refine his techniques, making his counterfeit currency increasingly indistinguishable from the real thing. This back-and-forth dance continues, with the counterfeiter getting better at producing convincing fakes and the detective getting sharper at detecting them.

When considering the vast landscape of artificial intelligence, the coexistence and convergence of discriminative and generative AI models represent one of the most exciting frontiers. One such exemplar of this harmony is Generative Adversarial Networks (GANs). As the name suggests, GANs pit generative and discriminative models against each other, resulting in a powerful system that can create, refine, analyze, and invent.

Now, replace the counterfeiter with a generative model and the detective with a discriminative model, and you have the essence of a GAN or a Generative Adversarial Network. The generative model creates (forgeries), and the discriminative model discerns (detects). Their dynamic rivalry propels them to improve, creating highly realistic outputs, whether they are images, text, or something else.

The power of GANs lies in the competition and cooperation between these two elements. As the generative model gets better at producing convincing content, the discriminative model must improve its ability to discern the real from the fake. This ongoing adversarial relationship leads to both models' continuous refinement and improvement.

So, what does this mean for practical applications? The promise of GANs is nearly as broad and varied as the field of AI itself. They have been used to generate realistic human faces, translate images from winter to summer, create 3D models from 2D images, and much more. In art and entertainment, GANs can be used to create original art pieces, design clothing, and generate unique character models for video games. In healthcare, GANs have the potential to generate synthetic medical data, which is useful for research in areas where data collection can be challenging due to privacy concerns. They could also create detailed 3D models of organs to assist doctors in surgery planning. As GANs evolve and improve, their potential applications will only broaden. By unifying the forces of generative and discriminative models, GANs unlock a new dimension of machine intelligence.

Before we dive into why AI has taken center stage in our everyday conversations and why it seems we cannot go a single day without hearing mention of it, let's take a slight detour. It's important to untangle the web of technical terms that have entered our regular vocabulary through metaphors and make them more digestible for everyone. You can skip to the last lines of this chapter if the technical jargon does not evoke your curiosity.

Decoding The Technical Jargon of The AI World

Let's start with the metropolis of "Discriminative AI". In the city's bustling heart, there was a busy construction site where an elegant building was taking shape. This building was no ordinary one. It was a "Model" - a unique shop with the power to transform information into knowledge. "Data points "into the relevant "Labels"

Our "Model" shop is designed and constructed following the genius vision of a skilled human - the "Architect." This Architect creates the "guiding principles or procedures," like blueprints for our building. These principles outline what the shop should look like, how it should function, and how it should be constructed. In the world of AI, we call these principles "Algorithms."

Then we have our constructer, the "Engineer". This Engineer is not a regular human but a computer system akin to a meticulous robotic engineer. The Engineer takes the Architect's guidelines and the Algorithm and starts constructing the shop, our Model. But how does the Engineer know what to build? That's where the "Data" and "Labels" come into play.

Imagine the Data as a collection of objects. It can be a collection of images, a text corpus, etc. Each data point is different, just like different building materials have distinct characteristics. The Labels, however, are the identifiers or tags associated with these objects. They explain each type of data point and connect them with semantic understanding shared with humans. E.g., "Persian Cat," "Labrador dog," "Sunny beach," "Snow Mountain," etc.

The process of building our model, or shop, using the Algorithm to grind the data and labels correctly and construct the shop is what we call "Training." This Training can take a long time and a lot of computational energy, much like it takes weeks and months to construct a physical building.

The data used to construct our model shop will determine its unique characteristics, like how different materials and designs result in different types of shops. A Model built using images of animals and their corresponding labels would be different from a Model built with images of human faces and their names, much like how a grocery store differs from a clothing boutique, even though the same Architect and Engineer built them.

Finally, when the Model is finished, we're ready to open the shop for business. New data can enter this shop, stroll around, and come out with an appropriate label. This process is known as "Prediction or Classification". It's fascinating, isn't it? How our Model can take a piece of raw, unprocessed data and attach to it the perfect label, like a shop delivering the perfect product to a customer!

Now, imagine our bustling city is growing. As the needs of its residents become more diverse, the variety of shops they require also increases. Each new shop or Model we wish to build must meet the demands of its target audience's "Target Market." The sophistication of these shops is determined by the time, budget, and specifications provided by the client.

Consider a high-end clothing boutique catering to an elite clientele seeking the best in fashion. To satisfy their needs, the client invests more time and money into gathering a large volume of quality raw materials - our Data. These materials are meticulously sorted and curated with clear Labels to ensure the finest level of detail. Using these materials, the Engineer constructs a grand, intricately designed boutique based on the Architect's algorithms. This boutique, our advanced Model, will then be capable of delivering high-quality predictions perfectly tailored to the needs of the elite clientele. However, these sophisticated predictions take more time and come at a higher cost, much like the designer clothes in the boutique.

On the other hand, imagine a budget clothing store catering to a different audience. This audience values speed and affordability over high-quality predictions. The client may likely invest less time and money in data collection and curation for this target market. Thus, the raw construction materials are more straightforward. The Engineer, following the Architect's instructions, can quickly construct a simpler shop. This shop, or Model, can provide faster but lower-quality predictions. This is akin to the budget clothing store offering affordable, off-the-rack options for its customers.

In both scenarios, the Architect - the human genius behind the scenes - has a crucial role. They equip engineers with the Algorithms and tools necessary to cater to diverse target markets. Depending on the application's needs, they empower the Engineer to construct everything from the most sophisticated boutiques to simple budget stores.

In the same way, in deep neural networks, the core algorithms can be customized for specific needs. Based on the end application, they can be scaled up and down for both speed and accuracy.

Now, let's imagine a grander scale, a wider array of needs from our city residents, ranging from everyday household items, furniture, and clothing to even exotic luxury goods. Our client wants to create something beyond a boutique or a simple shop to cater to this enormous demand. They want to build a "Shopping Mall" where everyone can find their needs.

To do this, the client sources an extensive and varied collection of materials or 'Data.' This enormous, diverse collection, called "Big Data," encompasses many categories and varieties. It's akin to sourcing materials from around the globe, considering every possible requirement that customers might have.

The Engineer, our trusty Algorithm, takes this Big Data and, using our Architect's blueprints, begins building this 'Shopping Mall.' This construction phase is a complex and lengthy process, requiring significant time and resources.

When this intricate Mall, our sophisticated Model, is finally ready, it's a sight to behold. It can cater to anyone and everyone, allowing them to walk in with their unique requirements and walk out with their perfect predictions. Whether identifying a cute tabby cat, recognizing an abstract painting, or even discerning the different species of birds, the Mall has a place for every need. These are public models like ChatGPT or Gemini created by big tech companies.

Of course, the convenience and range of services offered by the Mall come with a premium. However, the cost is justified by the convenience and accuracy of its predictions. Customers can get everything they need in one place, much like how these massive AI models handle a wide variety of data and deliver high-quality predictions.

Architects, the pioneers in AI, are constantly innovating. They work tirelessly to design blueprints for more extensive and accurate Models. At the same time, they strive to make these models more efficient and affordable in hopes that these 'Malls' can be replicated across multiple cities (or applications), bringing the power of AI to everyone, everywhere.

The "Mall" stands ready, waiting for the next piece of data to walk in and find its perfect label.

In our city of AI, the quality of the construction greatly depends on the quality of the raw materials – the data and their associated labels. Having the right label for each piece of data is crucial. It's like ensuring each type of material is correctly identified and placed in the right section of the building during construction. This construction method, where the data point and its corresponding label are provided to the Engineer or the Algorithm, is called "Supervised Learning."

Here, the client provides the data points and their specific labels to the Engineer for the construction. However, gathering vast amounts of data and manually attaching the right labels can be daunting for clients. It's like sorting and tagging every brick, beam, and window before construction begins. While this approach often results in the highest quality predictions, the manual labor required to label each data point can be substantial and time-consuming.

Recognizing this, architects and innovators developed new construction methods. In this new method, "Unsupervised Learning," the raw materials, or data points, can be provided to the Algorithm without explicitly labeling each item. In pursuit of efficiency, our ingenious Architects devised this solution. In this method, the data points automatically find their groups or 'categories.'

"Unsupervised Learning" is like supplying a heap of assorted bricks to the Engineer and letting them sort them into different categories based on size, shape, or color. Once the Engineer finishes the sorting, a human Architect assigns names to these categories. For example, all the red bricks could be categorized as 'for walls' and all the white bricks as 'for interiors.' Imagine if all the construction materials - the bricks, beams, glasses, everything - are dumped at the construction site without labels. The Engineer, our Algorithm, can recognize patterns and similarities amongst the different materials. It can group all similar materials, say all the bricks into "Category 1", all the beams into "Category 2", and so forth. At the end of this grouping process, the Architect can simply label each category - "Category 1 as bricks", "Category 2 as beams", and so on.

The advantage of this approach is that it significantly reduces the need for manual labor, as the need for explicit labeling for each data point is eliminated. It's like sorting the bricks without inspecting and labeling each one individually. This process greatly simplifies the preparation phase and saves considerable time and energy. However, this convenience comes with a trade-off. The predictions from Unsupervised Learning models might not be as accurate as those built using Supervised Learning. It's like constructing a building without knowing the exact function of each component. The overall structure might not be as perfect as one might desire. This is the consequence of the system being unable to perfectly categorize the data points every time, akin to the Engineer sorting bricks based on color. At the same time, their strength was a crucial characteristic.

But fear not; our Architects are not ones to rest on their laurels. They are continually striving to bridge this gap in accuracy. They are exploring innovative techniques to enhance the predictive power of Unsupervised Learning models.

In the heart of our AI city, where Data enters and leaves with new Labels, there's an underlying aspect that we still need to explore fully - the diversity of the Data. Imagine that the materials collected for constructing our 'Prediction Super Mall' have a peculiar trait. The materials - or Data - sourced for construction are specific to certain types of cats or dogs that are prevalent only in the USA. What happens if a customer brings in an image of a cat breed from Japan?

Our engineer or Algorithm constructed the mall based on the provided materials. However, when a customer comes with a data type unseen during the construction, the Mall might struggle to provide an accurate prediction. It's like a customer asking for a product that the Mall hasn't stocked because it wasn't aware such a product existed. The result? A customer walks away with a wrong prediction or product while trusting the Mall to be accurate.

This issue is what we call "Bias" in AI systems. It's an unconscious lean towards certain types of data due to the nature of the training dataset. The Prediction Mall, built using data from only one part of the world, may not be well equipped to handle data from other regions. This simple example can be extended to "Bias" in various forms in the real world.

The conscious or unconscious biases of humans when creating and curating datasets can significantly affect the outputs of these AI systems. It can cause them to deliver less accurate predictions when presented with data they are not familiar with. The essence of Bias in AI, as we've discussed, is a systematic error that can appear in the predictions made by the AI models. Bias can impact the performance of AI models in multiple ways, and it's often more pronounced when the model encounters data that differ significantly from its training set. Let's consider some real-world examples to illustrate this.

First, let's think about **facial recognition technology**. Suppose an AI model was trained predominantly on images of the faces of people of a certain age group, say young adults. This training data, therefore, lacks diversity in age. Suppose this model recognizes the face of an elderly person or a child. In that case, it might struggle to do so accurately. This is an example of Age Bias.

Similarly, suppose a model is trained mostly on images of faces from a specific ethnic group. In that case, it may not perform well on faces from other ethnic groups. This Ethnicity Bias can lead to significant inaccuracies. An instance of this was seen in 2015 when Google Photos, which used AI to tag and sort photos, mistakenly tagged photos of black people as gorillas. This was a case where the training data regarding skin color was not diverse enough.

Another example can be seen in language translation models. Suppose these models are predominantly trained on formal or literary texts. In that case, they might struggle to translate informal or colloquial language accurately, introducing a Language Style Bias.

Bias can also surface in job recruitment AI systems. Suppose an AI model is trained on historical hiring data of a company that has predominantly hired men for certain roles. In that case, the model might favor male candidates for those roles in the future, leading to Gender Bias.

These examples underline the importance of awareness of potential biases in training data. AI models are only as good as the data they learn from. Suppose the data lack diversity or are skewed towards particular characteristics. In that case, the models may fail to deliver accurate or fair results.

It's essential for both the Clients who own the Mall and the Target audience to be aware of these potential biases. Before utilizing the predictions from the AI system, it's important to understand the context in which the Model was built, including the data used for its construction.

In our ever-evolving city of AI, where Architects and Engineers tirelessly innovate and create, there's a constant hum of new ideas and terminologies. One can often hear the letters' CNN,' 'RNN,' and 'Transformers' being thrown around in discussions. These abbreviations may seem complex, but they're simply different types of blueprints for constructing our AI shops or 'models.'

Let's consider CNN or Convolutional Neural Network. Just as an architect may specialize in designing layouts for art galleries to best display paintings and sculptures, a CNN is specially designed to handle visual data, like images. It has a unique ability to scan an image in small parts and understand the whole picture.

Then we have RNNs or Recurrent Neural Networks. Imagine a shop where the order customers visit is just as important as that of the customers. That's precisely the expertise of an RNN. It's designed to understand and predict sequences, like the words in a sentence or the notes in a piece of music. It remembers what happened earlier in a sequence and uses that information to understand what comes next.

But in our fast-paced AI city, today's groundbreaking innovation can become yesterday's news. Enter Transformers. These are like the skyscrapers of our city, towering over the previous models in terms of capability and flexibility. These revolutionary models were introduced by engineers at Google in 2017. Since then, they've been the blueprint for the city's most impressive structures.

You may have heard of GPT, BERT, or Llama. These aren't exotic pets but different variations of Transformer models. Each one has its specialty. For example, GPT is adept at generating human-like text. At the same time, BERT is a whizz at understanding the context of words in a sentence. Just as our city's architecture continues to evolve, introducing new styles and techniques, so does the landscape of AI. Each new model brings a fresh perspective, helping us tackle complex problems and uncover new insights.

Within the thriving metropolis of AI, a fresh challenge emerges - constructing a new Shop or Mall when raw materials or data are scarce. Some applications or Clients don't have the luxury of enormous data. Think of it as needing to construct a new specialty boutique store. Still, a limited amount of unique building materials are available.

This scenario could happen in specialized fields like medical imaging, where we might not have a vast library of MRI or CT scans of human lungs as we do for pictures of cats and dogs.

Faced with such a problem, our diligent Architects and Engineers have devised a clever solution called "Transfer Learning." Think of it as repurposing construction materials from a pre-built structure to erect a new one.

In our city of AI, Transfer Learning works similarly. Instead of constructing a new shop from scratch, we take a pre-constructed shop (or a model) that might have been built for a completely different purpose, such as recognizing cats and dogs. We clean out parts of it, removing irrelevant sections and keeping the fundamental structure intact. Think of it as buying an old shop, clearing some parts of it, and remodeling it to suit the needs of the new boutique store.

Our Engineer, the Algorithm, then remodels this pre-constructed shop using a small amount of specialized target data, like MRI scan images. We reinforce the walls, add new sections, and repaint the interiors to suit the needs of the new application. This way, we managed to construct a boutique store capable of catering to the specific needs of detecting cancers or tumors, even when the available data was limited.

Through this process, even though we started with a shop that was once filled with images of cats and dogs, we now have a specialized medical imaging shop that can detect cancer or tumors from MRI scans. We've created a functional, efficient shop in a domain where only less data is available, leveraging the knowledge from one domain to aid in another.

Transfer Learning is a testament to the resourcefulness and adaptability within our AI city.

In the bustling AI cityscape, a new situation arises. Imagine a particular client whose business revolves around rare breeds of cats and dogs - ones like Ragamuffin or Burmese cats, or a Chinook or an Otterhound. These breeds are so rare that their images didn't figure prominently (or at all) in the initial mega data collection that went into constructing our Prediction Malls. Yet, this client has a unique dataset containing images of these rare breeds.

In response to this unique situation, our adaptable Architects and Engineers have a solution - a principle called "Fine Tuning." It's akin to remodeling an existing building for a new purpose, but the changes are more subtle and specific this time.

Consider this - the client has a unit in the sprawling Prediction Mall, but it doesn't cater to the unique needs of their business. Here's where fine-tuning steps in. The engineers slightly tweaked the existing shop using the client's unique dataset. They might modify the display, rearrange some shelves, or add unique signs to make it better suited to recognize and cater to the rare breeds.

Essentially, the shop is not being reconstructed but rather 'fine-tuned.' The transformation might seem cosmetic, but it's crucial for the client's business needs. This process enables the shop to provide accurate predictions for rare breeds - an application previously outside the Prediction Mall's purview.

The principle of fine-tuning embodies the versatility and adaptability that are hallmarks of our AI city. It's yet another tool in our AI toolkit that allows us to cater to an array of unique needs and applications, demonstrating the far-reaching potential of our AI city.

In the dynamic panorama of our AI city, Architects and Engineers are never content with the status quo. Their constant pursuit of innovation and improvement led to a revolutionary idea that would flip the dynamics of the city—instead of feeding the shops with data and receiving labels, what if we fed them with labels, and they could generate the data?

Imagine walking into a Shop, but instead of providing an image and asking for a label, you provide a label and ask for an image. This is not just any shop but a magical, transformative 'Super Mall' of AI - a generative AI system.

This cutting-edge concept turned the conventional use of AI on its head. It's like walking into the Shop, asking for a 'Persian cat,' and walking out with a fresh, unique image of a Persian cat. You provided a 'prompt,' and the shop generated the data. This is the essence of 'Generative AI'.

Of course, this seems simple, but there's much more complexity beneath. Generative AI's magic lies in its ability to interpret prompts and generate accurate and diverse data points.

Consider a complex prompt: "Generate an image of a Persian cat sitting on a garden table facing a lake on a sunny afternoon." Now, the AI doesn't just need to know what a Persian cat looks like and how it would appear sitting on a garden table; it also needs to know the visual aesthetics of a lake and the ambiance of a sunny afternoon.

But Generative AI doesn't stop there. It also needs to understand the prompt's intricate grammar and structure and visualize how the elements would merge into a singular, coherent scene. This complex dance of understanding, interpreting, and generating is an intricate ballet the AI system performs.

This revolutionizing shift, the ability to 'invert' the model and generate data from prompts, has broadened the horizons of AI like never before. It's not just about recognizing and labeling anymore - now, we're also prompting and creating. This has fired up massive interest in the AI landscape, taking it from a supporting role of analysis and prediction to a leading role of creation and generation. This is the new era of Generative AI, opening up many applications and possibilities and turning our AI city into a creative, generative powerhouse.

As AI systems' "Super Malls" grew more sophisticated, they exhibited new capabilities and unforeseen challenges. A particularly interesting and perplexing characteristic emerged when these models encountered unfamiliar data or prompts. This phenomenon became known as "Hallucination."

Imagine walking into a shop exclusively designed and built to recognize and categorize geometric shapes: triangles, squares, and circles. When presented with an image of a star, a shape it has never been trained on, the shop, or in this case, the AI model, does its best to relate this new, unfamiliar shape to what it knows. It might generate a set of probabilities like triangle 0.2, square 0.3, or circle 0.5 - effectively saying that it's most likely a circle. Still, it could also be a square or triangle.

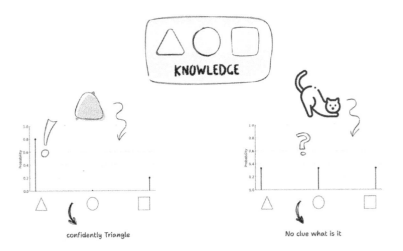

confidently Triangle   No clue what is it

However, suppose you were to provide this model with an image of a cat, which bears no resemblance to the shapes it knows. In that case, it might equally distribute probabilities among the familiar labels, giving something like triangle: 0.33, square: 0.33, circle: 0.33. It's essentially at a loss, and if pushed to give a single label, it may spit out a random label each time.

This scenario is more complex in a "Generative AI" world. The AI system has been fed vast amounts of text on history. If it's then given a prompt asking for an explanation of how quantum computers work, a topic completely outside its training realm, it might flounder. It might cobble together a response that looks and sounds plausible, even grammatically correct, but is fundamentally incorrect or nonsensical in its content.

This is akin to an AI system 'hallucinating' - creating something that seems real or meaningful but is based on a misunderstanding or lack of knowledge. It's as though the AI is attempting to construct a reality based on incomplete or irrelevant information, much like a hallucination in a human brain.

The following passage illustrates "AI Hallucination."

In the epoch of ancient civilizations, we had the Roman Empire, known for its grandeur, efficient legal system, and incredible engineering feats. If we were to equate this empire to classical computing, it's because of its deterministic nature. Just as the Romans had a strict code of laws and a clearly defined hierarchy, classical computers have binary states: 0s and 1s. Everything operates in a well-ordered, predictable manner.

But then came quantum computing, which we could liken to the mysterious civilization of the ancient Druids. Unlike the Roman Empire, the Druids were famous for their mysticism and their connection with the nature around them. These elements lay beyond the clear-cut and the definitive.Similarly, quantum computing doesn't confine itself to the rigid 0s and 1s. Quantum bits, or 'qubits,' can exist in multiple states simultaneously, thanks to a property known as 'superposition.' This is akin to how a Druid could be a healer, a lawkeeper, and a spiritual guide all at once. Just as the secretive and enigmatic ways of the Druids represented a completely different paradigm to the Romans, so did quantum computing present an entirely new way of processing information compared to classical computing.

This is especially crucial when considering public-facing AI systems, which may be asked to generate information on various subjects. If an AI system hallucinates, an uninformed user might be led astray by plausible-sounding gibberish, believing it to be accurate because it comes from an AI system.

Thus, while the AI city is expanding and innovating, it's also important to tread carefully. Just like in any city, some shops may offer things that look appealing but are ultimately of little value or even misleading. Vigilance and understanding are key to navigating this rapidly evolving landscape.

So next time you hear someone talk about AI, think of this bustling city with its myriad of unique, fascinating shops, each serving a different purpose and each built with a different set of materials. That's the magic of AI, simplified for you.

As we close this chapter, it's essential to note that the "Discriminative AI" stage has been growing and refining its act for years, significantly impacting numerous real-world applications. However, the spotlight has recently shifted to the rising star of the AI world - "Generative AI." Its recent breakthroughs have created a stir of excitement and curiosity that's impossible to ignore.

For the sake of simplicity and as we journey further into the realms of this book, whenever we mention 'AI,' it will be about this shining star, the "Generative AI." Consider this a little narrative shorthand, a simplified term for us to engage with the complexities of this fascinating, innovative field. So, let's turn the page and delve into the enthralling world of Generative AI.

3

UNMASKING AI:
WHY THE FUSS AND WHAT CAN IT
REALLY DO?

The artificial intelligence (AI) world is abuzz with enthusiasm, optimism, and a healthy dose of skepticism for some. It's not the first time we've been here. We've witnessed exhilarating hype cycles where AI's potential seemed on the brink of changing everything. Yet, we've also weathered the AI winters, periods of disillusionment where progress and funding seemed to freeze over, and cynicism chilled the enthusiasm of even the most ardent believers. Today, there is an excitement that feels palpably different. In the air, there's a sizzle, a spark, an electrifying energy distinctly unlike previous fervors. Yet the question on everyone's minds remains the same: Is this just another cycle? Another peak before the inevitable plummet into disillusionment and stagnation? Or, does this exhilarating chapter signify an inflection point, a pivotal turn, that will irrevocably alter the trajectory of our collective future? Let's dive into a fascinating story where the answers should become clear; for every shift in perspective, every new insight gained will not just help us comprehend the present, but also shape the way we envision and construct the future.

Linguistic Marvel: Bedrock of Human Progress

The quest to enable machines to comprehend and converse in human language - Natural Language Understanding (NLU) - has been a golden fleece for AI researchers. Natural Language Understanding, or NLU, is far from novel. Researchers and innovators have toiled in this field for decades. This pursuit is not a small feat.

The diversity of expressions, the subtlety of language, the nuances of sentiment, the context-dependent interpretations, the dynamic slang, and cultural dialects presented a puzzle of extraordinary complexity.

For a long time, our attempts at NLU were as rudimentary as the primitive grunts of our cave-dwelling ancestors. Individual AI systems must be meticulously trained with vast data for each specific task. The process was laborious, rigid, and had its limitations. The concept of Sentiment Analysis, classifying text by underlying emotions, was an early signpost of progress. Still, their versatility and accuracy were bounded by a narrow set of classification labels and lacked holistic comprehension.

Fast forward to our current epoch, and we are witnessing a revolution in NLU. The catalyst? Deep neural networks are modeled on the intricate workings of the human brain. Like a linguistic Big Bang, deep neural networks exploded onto the scene, expanding the boundaries of our AI universe. When applied to the realm of language, they led to the genesis of Large Language Models (LLMs). These models, trained on an inconceivably large corpus of data, manifested a unified framework for natural language understanding.

Like a symphony conductor flawlessly orchestrating a complex musical piece, these models can discern the delicate interplay of words and their contextual combinations. They navigate the nuances of words and phrases, their contexts and connotations, with an elegance and complexity that was once the sole domain of human minds.

The leap forward was monumental. LLMs could be used across various applications without requiring specific task training. But the magnum opus was their ability to understand and generate text. They now go beyond mere understanding - they create.

With a flair for mimicking various tones, styles, and content, these AI models emulated a congregation of language and domain experts. They didn't merely interpret language; they breathed life into it, generating prose as fluid and nuanced as a human author.

But why does this major leap in NLU ability warrant such enthusiasm, such breathless excitement? The role of language in our transformation from prehistoric cave dwellers to sophisticated space explorers offers a profound insight into the potential and power of AI-driven language understanding capability of machines.

When we reflect on the term 'technology,' images of computers, machinery, digital networks, intricate algorithms, and sophisticated external tools often rush into our minds.

We visualize the expanse of human ingenuity in designing devices that shape and advance our civilization. Yet, this one-dimensional definition overlooks the most primitive and fundamental technology: Language. Yes, "language" is a technology so seamlessly integrated into our daily lives that its technological aspect is often overlooked.

One must first dismantle conventional notions of technology primarily associated with material inventions to recognize language as a technology. Instead, let's envision technology as the embodiment of tools and systems devised to enhance our existence and propel us toward our desired objectives with greater efficacy. Within this broader perspective, language unfurls as an ancient tool—a technological marvel—that has been the cornerstone of all human progression.

Humans are not the only species capable of developing and using tools; various animals demonstrate tool-use behavior, from primates using sticks to extract termites to birds employing twigs for nest-building.

Yet, our species stands alone, not because of our ability to utilize the physical world but rather our ability to navigate the abstract. Language enables us to frame, name, share, and change reality. It grants us the power to encapsulate complex concepts and emotions into audible symbols, to transmit knowledge across generations, and to orchestrate large-scale social coordination.

Consider for a moment the scenario of early human communities. The human ability to communicate, to weave tales of hunts, to share the locations of ripe fruit trees, and to warn about lurking predators allowed survival knowledge to be shared and improved upon. The hunter's spear became sharper, and the gatherer's route became more efficient, not simply through individual trial and error but by sharing collective wisdom.

Language enabled the building and refining of communal knowledge repositories, marking the first step toward civilizations: communities at scale.

Imagine how different our world would be if the story of humanity were devoid of language. Without this unifying tool, we might have remained isolated tribes, never achieving the shared endeavor and collaboration that gave rise to agriculture, written script, mathematics, and all subsequent technologies.

The Power of Abstraction - From Symbol to Syntax

Language, as we perceive it today, is a marvel of abstraction. It's a sophisticated tool capable of condensing vast swaths of information into compact, shareable packages. This capability alone puts humans in a league of their own, distinguishing us from other species that, despite exhibiting varying degrees of communication, cannot encapsulate complex concepts and emotions into symbolic, verbal constructs.

Imagine a world without words, where communication is limited to gestures, physical actions, or rudimentary sounds. Consider the communication methods of our non-human counterparts in the animal kingdom. Many species can communicate, but their lexicon is typically bound by immediate, tangible realities, limited to actions or symbols. For instance, a monkey may convey danger or food availability through a particular gesture or vocalization, but the depth of expression ends there. Its communication remains tied to the present moment and the physical world.

Human language, on the other hand, transcends this immediacy. It creates a shared reality, a collective consciousness that extends beyond the 'here' and 'now.' It allows us to explore abstract notions, discuss past events, imagine and plan for the future, and share subjective experiences. It is a potent tool for cooperation and coordination, enabling us to work together on an entirely different scale.

The evolution from simple sounds to spoken words was a game-changer. It birthed a common encoding system that could be shared among large groups, propelling information exchange to a new level. This was the first major step in our linguistic evolution - the advent of speech.

The progression did not stop there, however. The next major leap came with the invention of writing systems. Our ancestors began to represent these spoken words with visual symbols, further expanding the reach and permanence of our communication. From simple pictograms mimicking the physical world, like those found in ancient cave paintings, they moved to abstract symbols representing sounds or concepts, giving birth to alphabets.

While spoken language broke barriers, the advent of written language bulldozed them. By transcribing spoken language, our ancestors could record knowledge, tell stories, and communicate across time and space. Alphabets - the smallest set of symbols that could construct an entire vocabulary - represented a significant advance in language technology. With this system, we no longer had to draw a lion to express the concept of 'lion.' A few pen strokes could convey the same meaning, allowing us to communicate more efficiently, with greater nuance, and over vast distances.

Across history and even today, we can see the profound effects of these linguistic milestones. Societies with spoken language had more complex social structures than those without. Societies with written language are more advanced in technology, governance, and civilization.

The Language of Machines

As we delve into the modern landscape of technology, we uncover a fascinating parallel between the development of deep neural networks (DNNs) and the evolution of human language. This parallel lies within the central principle that gives both their power: the concept of abstraction.

Human language, as we have seen, can abstract complex notions into individual words. But the real beauty of language lies in its capability to build upon these individual abstractions, layer upon layer, to construct elaborate edifices of thought. Single words combine to form sentences, sentences join to create paragraphs, and these paragraphs collectively weave vast tapestries of ideas, stories, and knowledge. This multilevel abstraction is a key feature of human cognition, allowing us to comprehend, communicate, and interact with the world in nuanced ways.

The versatility and expressive power of language lie in its combinatorial nature. When rearranged, the same set of words can create wildly different narratives. Let's illustrate with a simple example.

Paragraph 1:

On a quiet night, a cat lurked in the shadows under the shining moon. It was watching, fear evident in its wide eyes, as a dog gave chase down the empty street. The once peaceful scene, lit only by the moonlight and the dim glow from distant houses, was now fractured by the echo of a terrified yowl and claws scraping against asphalt.

Paragraph 2:

In the quiet corner of the street, where the houses meet, children often play under the watchful eyes of the moon. One night, their laughter filled the air, pushing back the shadows and filling the silence. A ball would roll, and a dog, no stranger to their joy, would chase after it. Nearby, a cat, undisturbed by the delight, was watching it all, the moon's glow highlighting its amused curiosity.

These two paragraphs use the same words but conjure completely different images and emotions. One scene is filled with fear and urgency, while the other brims with joy and serenity. This is the true power of language: the ability to form vastly different narratives with a small alphabet set. This flexibility, this capacity for creating infinite meanings, is what makes language an irreplaceable technology, not just for human communication but also for shaping our understanding of the world around us.

> **Similarly, the strength of DNNs stems from their capacity for abstraction at multiple levels.**

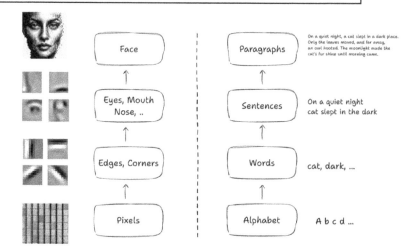

In DNNs, the process begins with extracting raw pixel data into primitive shapes. Each pixel, a mere point of color and light, holds little meaning. However, when these pixels are grouped, they form lines, edges, and corners - the basic building blocks of visual recognition.

The network then combines these shapes to create mid-level features, such as patches representing eyes, a nose, a mouth, or ears. It's akin to moving from words to sentences in our language analogy - these are complex entities formed by combining simpler elements.

Finally, at the highest level of abstraction, the network integrates these facial feature patches into comprehensive representations of

THE INEVITABLE AI

individual faces. This is equivalent to forming a coherent story or a book in language, an interconnected network of concepts that encapsulates a complex idea.

This multi-layered process of abstraction gives DNNs their uncanny ability to recognize faces from an array of pixels or understand natural language from a series of characters. It also makes DNNs surprisingly analogous to our human cognitive and linguistic processes.

This significant connection and capability is far from a mere academic curiosity. It's a testament to the power of abstraction, a principle that underpins both the success of our species and the latest advancements in artificial intelligence. It also shows how deeply our technological creations are intertwined with our natural cognitive capacities. It reveals why these advancements would not be a fad but a companion for the long-term future.

The Power of Universal Modality

As we immerse ourselves deeper in the currents of AI's ongoing revolution, the second key driver reveals itself - the broad applicability and inherent versatility of deep neural networks (DNNs).

Consider DNNs a universal toolkit, a common language of learning and understanding that transcends the boundaries of domains and data types. This is not merely a step forward; it's akin to discovering a Rosetta Stone for the AI world, an enabler of cross-domain communication and innovation.

At the heart of this universal framework lies the principle of symbolic abstraction and hierarchical combination of abstracted features, a concept that mirrors the very essence of human cognition.

Just as we navigate the world around us through the lens of layered abstractions - perceiving, comprehending, and interacting with an intricate reality - DNNs apply this same principle across many data domains.

This universal principle propels AI's capabilities to new heights, allowing the same base learning framework for processing different data forms, including images, text, audio, video, software, etc.

Historically, dealing with different data streams required creating specialized, hand-crafted features. Text was processed differently than images, and audio required separate algorithms. Combining these disparate algorithms was cumbersome, inelegant, and often ineffective. However, the dawn of deep learning heralded a sea change in this approach. Now, we can mix and match generative and discriminative components of algorithms and integrate different data domains seamlessly.

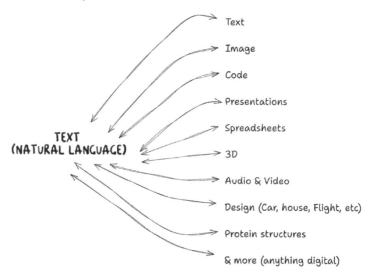

The transformative capacity of this technology opens up an expansive canvas for creativity and innovation. We can perform feats that seemed improbable a few years ago, like generating a piece of music inspired by an input image or turning a written text into a visual scene. These intriguing possibilities, while demonstrating the potential of multimodal learning, are just the tip of the iceberg.

And what's more, the complexity of these networks can be meticulously calibrated. Whether it's a simplistic network for a relatively straightforward task or a sophisticated, layered behemoth capable of deep understanding and complex decision-making, DNNs can be tailored to meet diverse needs. This scalability is equally crucial when considering the processing speeds and power needed for different tasks. From the expansive power of cloud-based systems to the lean efficiency of edge computing, the same core algorithms can be utilized across different scales and architectures.

The unlocking of this core learning mechanism, fueled by the abundance of data at our disposal, represents a significant shift in our AI journey. Now, the potential applications and data domains to which DNNs can be applied are only constrained by the bounds of human creativity.

When we ponder the complexity of our world and the myriad of creations brought forth by human intellect and endeavor, we must acknowledge the single most significant tool that has made this possible: language. It serves as the gateway to everything humankind has ever produced, from the grandest theories of the cosmos to the simplest expressions of joy or sorrow. More than a mere tool, language is the very interface through which our minds connect, creating a vast, intertwined network of human knowledge.

It is the backbone of our collective intellect, allowing a single idea to ripple through a community, amplifying and evolving as it travels, often blossoming into magnificent innovations over generations. Language is the conduit through which our internal cognitive world connects with the external physical world. It enables us to disseminate our thoughts, to share our ideas, to influence, and be influenced. Consider the power of an idea. In isolation, its potential is limited by the confines of a single mind. But when shared through the medium of language, its power multiplies. It can travel vast distances, penetrating different minds, amassing, adapting, and augmenting with each new interaction. Like a pebble dropped into a pond, an idea set loose in a linguistic community creates ripples that spread and amplify far beyond its origin. This seamless flow of thought across individuals and generations enables the blossoming of ideas into philosophies, inventions, cultures, and civilizations.

Yet, the role of language transcends the sphere of communication; it forms the bedrock of our cognitive processes. Language is not just how we express our thoughts but how we shape them. It allows us to formulate abstract concepts, analyze and synthesize information, solve problems, and envision future possibilities. When we build a house, we begin with individual elements: bricks, mortar, timber. Likewise, when we construct an idea, we do so by assembling simpler, abstract concepts. And just as the house forms in the physical world, our complex ideas get shaped in the scaffold of language. It is the medium of our minds. As we ponder, plan, and perceive, we do so through the lens of language. It provides the symbols and syntax that structure our thought processes and, in doing so, shapes our cognition, our perspectives, and our decision-making. As the renowned linguist Benjamin Lee Whorf suggested, language structure can influence how we think and view the world, a concept known as linguistic relativity.

Language is also the medium of instruction, enabling the transference of knowledge and wisdom across generations. It is the medium of governance, allowing us to construct laws, regulations, and societal norms. It is the medium of innovation, facilitating the exchange of ideas that spur creativity and progress. It is the medium of emotions, allowing us to share joy, sorrow, fear, and love.

The ability to express and understand language has been integral to our survival and progress as a species. But the consequences of this ability extend far beyond the practical. Language shapes our perception of the world and our place within it. It enables the creation of art, literature, music - all the myriad ways we express our understanding of life and the human experience.

Language is the prime mover of all we have created and will create in the future, the catalyst of our triumphs and failures.

In all its richness and versatility, language is also the vehicle that carries the weight of our scientific inquiry and intellectual curiosity. From the intricacies of quantum physics to the vast canvas of astronomy, from the microscopic realm of biology to the abstract landscapes of mathematics, language remains the common thread that binds these diverse disciplines. Without language, science would be an inaccessible fortress with all its theories, hypotheses, and laws.

Consider the concept of gravity. This invisible force that governs the movements of planets, stars, and galaxies is abstract, intangible, and nearly impossible to comprehend without the benefit of language. How do we grasp the concept of an invisible force that pulls objects toward each other?

Through language, we can describe this phenomenon, breaking it down into its fundamental components—mass, distance, attraction—and weaving these threads into the fabric of a theory. This process of description, explanation, and theorization is at the heart of science and made possible by the tool of language.

Beyond being a medium for explaining and understanding, language also provides a platform for critical discourse. It is the arena in which ideas are tested, scrutinized, and refined. Through debate and disagreement, powered by language, we chisel raw, initial thoughts into polished theories and inventions. The crucible of scientific progress is as much about dissonance as harmony, as ideas battle, adapt, and evolve in the linguistic arena.

Returning from the detour of "Impact and importance of language," we arrive at a thrilling frontier of artificial intelligence: multimodal and intermodal learning. We will now realize why this parallel between "language development and Intermodal learning" is significant. This concept, harnessed by deep neural networks (DNNs), refers to the ability to process and correlate information across various types of data or modalities, such as text, images, audio, and video. The breakthroughs in this field, characterized by the flexibility to 'transform' between different data formats, signal an unprecedented leap in the potency of AI.

The most significant impact of multimodal learning is the integration of the 'text' modality with other forms of data. This combination allows AI systems to 'understand' and 'interpret' various data types in a context enriched by linguistic information. For instance, an AI model trained on image and text data can recognize a dog in a picture and generate a descriptive sentence such as "A black and white dog is playing in the park." The buzz is not just about machines' ability to calculate, analyze, or execute tasks— it's about their newfound ability to 'understand' and 'use' human language with all forms of digital data.

Let's take a moment to absorb and reflect on the potential of machines.

Machines As Partners in Creation

Prepare to redefine your understanding of the relationship between humans and machines, for we are looking at the dawn of a monumental shift. This latest breakthrough in AI is set to revolutionize our interaction with technology and forever alter the dynamics of our co-existence.

The transformative nature of artificial intelligence, particularly its proficiency in understanding and using natural human language, is not merely a progression—it's a revolution. Language, as we have delved into, is not merely a vehicle for human communication; it serves as the conduit for our thoughts, dreams, and ideas. Now, imagine the power of marrying this richness of language with the computational prowess of AI. The result? A seismic shift in the role of computers - from executors of pre-determined code and custodians of information to active partners in creation.

> **Simply put, we will interact with computers and all the diverse forms of processing they perform through natural language.**

In the age of AI, language has become more than a tool—it has become a bridge, connecting the human mind with the computational prowess of machines in an unprecedented manner.

By tapping into the richly expressive medium of language, the stronghold of human cognition, AI systems can now begin to comprehend and give form to our thoughts. They are evolving into intelligent entities that can grasp the subtleties and nuances of our expressions, allowing us to interact with them in ways we once could only imagine. This evolution elevates machines to a higher level, transforming them into partners in our creative processes.

Gone will be those days of interacting through the sterile clicks and navigations of pre-designed interfaces. Picture, instead, a world where we converse with machines as naturally as we would with a colleague, a friend, or even our children. With its strong understanding of our language, AI becomes an extension of our thoughts, providing form and substance to our ideas across many domains - from audio and video to software, genomics, and more. In these domains, where machines already excel in processing vast amounts of data, the infusion of linguistic understanding could create a synergy that propels our capabilities to new heights.

These recent advances in AI are akin to the advent of accessible personal computers. The personal computer revolution didn't just give us word processors and spreadsheets; it digitized our world. It allowed us to simulate weather patterns, design complex structures, compose music, and explore the mysteries of the cosmos. Similarly, the advent of machines that can understand and respond to natural language won't be limited to chatbots and digital assistants. They herald a future where language is the key that unlocks a dynamic interaction between humans and machines across myriad domains.

The implications are far-reaching and deeply profound. This powerful partnership could benefit every sector, industry, and human endeavor. Imagine designing a building by simply describing your vision or composing symphonies through a conversation about how music makes you feel. Envision a healthcare scenario where AI accurately translates a patient's subjective symptoms into precise medical terminology, aiding in quicker, more accurate diagnoses. The possibilities are endless and limited only by the bounds of human creativity.

The energy of this paradigm shift is palpable, and it's only the beginning. With each passing moment, new applications are imagined, new solutions are developed, and new boundaries are pushed.

Real-World Magic of Generative AI

As we've ventured through the philosophical depths and conceptual vistas of Generative AI, let's now shift our gaze to the tangible: its applications in the real world. We're not about to plunge into the nuts and bolts of specific software solutions; instead, we'll unfurl the panorama of possibilities these powerful systems present. Our journey begins with 'text-to-text' systems, the champions of which include the well-known 'ChatGPT.'

Imagine a massive library, with shelves reaching up towards infinity, teeming with all the public domain text the web offers. Billions of lines from blogs, tweets, LinkedIn posts, Reddit conversations, open-source books (fiction and non-fiction), Wikipedia articles - the list is staggering. This is the training ground for our large language models (LLMs). They are the diligent students of our digital universe, absorbing and understanding the vast spectrum of human thought, emotion, and knowledge expressed through text. The result is nothing short of magical.

So, what can you, as an individual user, expect from these digital wizards? What kind of alchemy can you create?

Consider crafting an engaging social media post. You have a seed idea, but the right words elude you. Enter the generative AI system. It takes your nascent thought and weaves it into a tapestry of compelling prose that resonates with your audience. Or perhaps you're seeking to express your feelings for a loved one through the timeless medium of poetry but find yourself tongue-tied by emotion. The AI system becomes your bard, crafting verses imbued with the sentiment you wish to express, personalized for the object of your affection.

From boosting your professional profile with a powerful resume and persuasive cover letters to brainstorming ideas for your next big project, AI systems are ready to assist. They can serve as dedicated learning partners, helping you explore and understand complex concepts interactively. They can effortlessly translate between languages, opening the world's treasury of knowledge to you. And they can paraphrase your core ideas into various formats, honing your message for different audiences.

Let's grasp this power and capability by looking at an illustration where you can write the same concept in multiple ways in seconds. The following example has been generated by using ChatGPT for illustration purposes only:

Let's take the concept of "The Earth orbiting the Sun." and write it in many different ways:

Formal Academic Style: The celestial body we inhabit, Earth, engages in an elliptical revolution around the star categorized as the Sun. This cyclical journey, dictated by gravitational forces, constitutes a fundamental aspect of our planetary system's architecture. This annual pilgrimage around the Sun establishes the basis for our calendrical systems and contributes significantly to the cyclical nature of seasons observed on Earth.

Casual, Conversational Tone: You know how it's summer one minute, and then before you know it, it's winter? That's all because our sweet home, Earth, is taking a year-long joyride around the Sun. Earth is on this giant cosmic carousel, courtesy of gravity, and the Sun's in the middle!

Poetic Expression: In the vast ballet of the cosmos, our Earth, a humble dancer, twirls in a year-long waltz around her luminous partner, the Sun. Guided by the unseen hands of gravity, she traces an elliptical path, a silent ode to the rhythms of time and the changing seasons.

Humorous Tone: Picture this: Earth, in a green dress, twirling around the Sun like a girl who's had too many solar martinis. She's got this crazy one-year dance card booked thanks to a burly bouncer named Gravity. So we get this funky changing room thing going on - swim trunks today, snow suits tomorrow. Who knew celestial mechanics could be such a riot?

Small Poem: In the theater of space so wide, Our Earth takes the Sun for a ride. A year-long dance, in space, they prance Under Gravity's unseen guide. She twirls, spins, and then swings in the ballet of cosmic things. With seasons four, who could ask for more? Oh, the joy this orbit brings!

Victorian Style: Indeed, dear interlocutor, our beloved terra firma doth partake in a grand promenade around the resplendent luminary known as the Sun. Under the invisible hand of gravitational forces, Earth partakes in this circuit over twelve calendar months, thereby governing the succession of our seasons.

Pirate Language: Arr matey! Ye know how our ship, Earth, be sailing 'round the bright star, the Sun? Aye, she makes a grand tour once every year, steered by the invisible winds of gravity. She does! That is why we are havin' seasons, so stow yer grog and heed the tale of this grand voyage!

To a Grandma: You know, dear, just like the way we have our daily routines, Mother Earth has her own too. Every day, she twirls around while she dances in a big circle around the Sun, taking a whole year to do it! That's why we get to enjoy different seasons throughout the year.

To a 5-year-old Kid: Imagine you are spinning around in a circle while walking around a big bonfire. The bonfire is the Sun, and you are like the Earth. It takes you one whole year to walk around the bonfire, and while you do this, sometimes you feel hot sold - those are our seasons!

Julius Caesar (Marc Antony) Style: Friends, Romans, countrymen, lend me your ears; I come to speak of Earth's voyage around the Sun. As surely as Rome's glory is unending, so does our world embark upon an eternal elliptical path around the Sun. Guided by the silent force of gravity, it completes this grand journey every year, bringing us the change of seasons. Just as Rome has times of war, peace, growth, and rest, the Earth has winter, spring, summer, and autumn. Remember, as the Earth continues its steady course, so must Rome stand resolute and unyielding.

And this is just the tip of the iceberg. The realm of text-based generative AI is as vast and varied as the World Wide Web itself. It's an exciting landscape with potential that is waiting to be explored. Let's look closer at what applications we will see in the future.

Beyond the wide-ranging applications we've explored, generative AI's power extends to some rather specific and often challenging tasks. Let's consider some of the intricate documents we encounter daily - contracts, insurance papers, or those infamous 'terms and conditions' we usually click 'agree' to without a second thought. Here's where AI truly shines, unweaving the labyrinthine language of legalese and transforming it into easily understandable terms.

Picture this scenario: you're about to sign a rental agreement, and the contract reads like a foreign language. A generative AI system could read and interpret the contract and communicate its findings clearly and understandably. What do those convoluted clauses mean for you as a tenant? Are there hidden charges or unfair provisions? The AI system equips you to make informed decisions by translating legalese into plain language.

Suppose you're a freelance designer and have just received a dense contract from a new client. Instead of pouring over the text line by line, you could simply input the document into an AI system. Initiating a chat, you might ask, "What are my obligations under this contract?" Drawing from its understanding of the document, the system responds with a list of your responsibilities and deliverables. You might follow up with, "What is the termination clause in the contract?" or "Are there any penalty clauses I should be aware of?" It's as easy as having a conversation.

This interactive method of reviewing documents saves you time. It ensures you're aware of all the important elements of the agreement. You can engage in more productive conversations with your clients and make informed decisions quickly and confidently.

Insurance documents are notoriously complex. They're packed with jargon, conditional statements, and subtle clauses. You're reviewing an insurance policy bristling with jargon and complex conditions. What's covered, what's not, and under what circumstances? Again, the AI system comes to your rescue, highlighting the essential points, identifying potential issues, and answering your questions interactively as if conversing with an expert. Imagine you've just bought a new health insurance policy and want to understand the specifics. You feed the document to the AI system and start your interactive session with "What are the key benefits of my policy?" The system gives you a summary. You continue with "What exclusions should I be aware of?" or "What is the process for filing a claim?" By conversing with the AI, you swiftly comprehend the nuances of your policy without having to sift through pages of complicated language. When did you last read the 'Terms of Use' or 'Privacy Policy' before installing a new app or signing up for an online service? With an AI system, you could quickly get answers to what data the service collects, how it uses your data, and what rights you have.

The opportunities don't end here. Imagine using AI to guide you

through complex scientific papers or technical reports, simplifying jargon, explaining concepts, and answering your questions on the fly. Consider using AI as a tutor for learning a new language, providing translations, explanations, and even interactive conversation practice. Or think about using AI for creative writing, offering suggestions, crafting compelling sentences, and helping you overcome writer's block. From mundane tasks to complex challenges, the potential applications of generative AI systems in our daily lives are numerous and diverse.

The intriguing possibilities we've explored are not just figments of a distant, speculative future. These AI capacities are present, tangible, and accessible for anyone to experiment with today. Now, however, let's rev up the engines of our imagination and take a thrilling journey into the near future. For a while, let's set aside the ethical, practical, and societal implications of AI's advancement and take a moment to marvel at the astonishing transformations that lie on the horizon. Rest assured, we will delve into the comprehensive examination of the challenges and considerations surrounding this technology in subsequent chapters.

The possibilities for AI in more professional scenarios are just as astounding. It can be a powerful ally in research and understanding complex public documents, transforming these traditionally laborious tasks into interactive and efficient experiences.

Research, especially academic, often involves combing through a vast body of literature. Traditionally, this involves keyword-based searches that may or may not yield relevant results. Imagine inputting your research question into an AI system that uses keywords and semantic search to scour existing literature. You could ask, "What are the latest advancements in quantum computing?" or "What are the key debates in climate change science?" Further, you can interactively explore the literature you have collected through AI. Queries like "What are the main conclusions of these papers?", "Where do these papers disagree on the impact?" or "What are the open questions in this field?" will help you gain a nuanced understanding of your topic quickly. This nuanced, comprehensive, and interactive literature review can save researchers countless hours and guide them toward more effective research paths.

Interacting with the government can often feel like navigating a labyrinth of complex regulations, varied departments, and interconnected documents. AI can be an invaluable guide in this context. Let's say you're an entrepreneur in the agriculture sector seeking to understand the government support available to you. The information you need is spread across various government websites, departments, and documents. It would take hours, if not days, to manually locate and decipher this information.

Enter generative AI. You could ask, "As an agricultural entrepreneur, what kinds of government support can I avail?" The AI system would fetch the relevant information from various sources and summarize and present it contextually. It then collates relevant information from tax benefits, subsidies, training programs, export support, insurance schemes, and more, tailor-made to your specific context. It could identify opportunities you weren't aware of, such as a new sustainability incentive or an emerging technology grant.

This use case dramatically simplifies complex tasks that would traditionally require specialist advisors. It democratizes access to information, making it readily available to everyone, irrespective of their resources or expertise.

Breaking New Grounds: AI Revolution in Law

The field of law is notorious for the amount of reading it requires. Attorneys, paralegals, and law students alike are all too familiar with the daunting piles of paperwork intrinsic to their profession. Lawsuits, legal codes, contracts, and past precedents are just some of the many document types that legal professionals have to grapple with daily. Lawyers must navigate through immense archives of law articles, analyze complex individual case documents, and dissect previous judgments. Moreover, they need to weave these strands of information into a compelling, logical, and legal narrative. The sheer time and intellectual rigor this process demands are immense. What if we could introduce an AI assistant into this equation?

Many documents are at the heart of every legal case: affidavits, depositions, contracts, emails, and more. Understanding and collating these into a coherent argument is a painstaking task. Generative AI could revolutionize this process. Picture a scenario where this is deployed in a law firm. With its immense capability to consume and analyze vast amounts of text in mere minutes, it would be like having a battalion of diligent junior lawyers tirelessly working around the clock. Lawyers could engage in a natural language conversation with the AI system, asking it to summarize complex case files, identify contradictions, extract key points, or even draw comparisons with other cases. This allows unprecedented interaction with the case data, fostering a deeper understanding and more comprehensive case preparation.

The AI could analyze individual case files, pore through law articles, examine previous judgments, and consolidate all this information into a logical and cohesive case brief. It could provide a comprehensive view of the case at hand, highlight similar cases from the past, outline possible legal arguments, and suggest strategic moves based on previous rulings. This would offer legal professionals the unprecedented advantage of scanning and synthesizing legal literature at breakneck speed, freeing their time to focus on strategic thinking, argumentation, and client consultation.

The legal world thrives on precedence. Previous court judgments often influence current cases, making extensive legal research necessary. Imagine inputting a query into an AI system like, "Show me cases where the 'stand your ground' law was successfully invoked." The AI could then scan through a vast library of past cases, legal articles, and judgments, returning a list of relevant cases and a nuanced analysis, identifying patterns, key arguments used, and the ultimate outcomes. This detailed, interactive exploration of legal precedents can enhance a lawyer's strategic arsenal, leading to more robust case presentations.

However, it's not just about efficiency and speed. At its core, this technology can potentially make legal services more accessible. For instance, AI could democratize access to legal advice, making it available to those who may not have the means to hire a lawyer.

AI & Healthcare: A Prescription for Progress

Let us now step into the realm of healthcare, an intricate blend of complex diagnostics, patient histories, treatment methodologies, and continuous advancements. Imagine the labyrinthine corridors of a hospital, the steady hum of life-saving machinery, and the ceaseless rhythm of footsteps, each belonging to someone dedicated to the service of others. This is where medicine meets human frailty, and every decision could make a difference between life and death. A doctor's job is, without a doubt, one of the most challenging. From recognizing patterns among a myriad of symptoms to understanding a patient's intricate medical history, from diagnosing the issue accurately to deciding on the most effective treatment method from a vast repertoire of possibilities - the responsibility is colossal. The introduction of generative AI could be an absolute game-changer within this arena.

Now, imagine if the doctor could have an AI-powered assistant, a virtual partner equipped with the power of Generative AI. This system, capable of consuming vast swathes of interconnected medical data, could function as a legion of junior medical assistants, working tirelessly in the background, sifting through medical histories, recognizing patterns, and presenting possibilities.

Doctors can accomplish far more when armed with an AI-powered tool that can access, analyze, and comprehend the labyrinth of medical literature, research studies, patient histories, and treatment methodologies. The AI assistant would not just be a passive bystander. It could proactively provide crucial insights, alert the doctor to potential complications, suggest differential diagnoses, and provide a comparative analysis of different treatment options.

The AI system could bring forward cases with similar symptom profiles, present potential diagnoses based on these matches, and propose various treatment options that have yielded the best outcomes in such cases. This could be done in real time, making the doctor's decision-making process much more informed, data-backed, and efficient.

The true marvel of such a system lies in its ability to learn from each interaction, each diagnosis, and each treatment. It can continuously update its knowledge base, refine its predictions, and offer increasingly accurate recommendations. This is like having the collective wisdom of hundreds of doctors, accumulated over years of practice, available at the touch of a button.

The AI Difference: From Information to Knowledge

A common question that can come up in our discussions about the potential of these systems is, "How is this different from our existing search-based technologies? Aren't we already able to find what we need using internet search engines?" While it's true that search engines have revolutionized our access to information, Generative AI takes this a step further by elevating our interactions from mere data retrieval to knowledge synthesis.

The quintessential difference lies in the depth and breadth of understanding these AI systems provide. Consider the search engine – an amazing tool in its own right, but essentially a 'pointer' to potential sources of information. It can guide you to websites or articles that may have the answers you seek. Still, it cannot directly provide you with a comprehensive response or perform complex tasks.

Generative AI systems, however, are like skilled craftspeople, capable of carving out coherent and detailed responses from the raw material of the internet's vast unstructured data forest. These systems are data-to-knowledge aggregators capable of understanding the nuances of the queries, deciphering the intent, and generating tailor-made solutions in real-time.

Think of current search technologies as vast, sprawling forests of information. They can point us toward a tree we seek. Still, we must make the arduous journey, navigate potential obstacles, and discern if the tree is needed.

On the other hand, large language models (LLMs) powered Generative AI systems act as intelligent guides through these forests. They do not merely show us the way to the tree; instead, they traverse the terrain on our behalf, understand the essence of what we're seeking, and return with the exact fruit we desire, ready for consumption.

To appreciate the difference, think about this: A search for a recipe on Google requires you to know exactly what you want to cook. A simple search query might yield millions of results, an overwhelming prospect for anyone searching for quick, meaningful answers.

However, with a Generative AI system, you can input a list of ingredients you have on hand, and it can generate unique recipes for you. The system can think in the abstract and even suggest a vegan substitute for a particular dish without requiring you to pore over multiple web pages. The difference is like receiving a list of ingredients versus being presented with a ready-to-eat meal.

Likewise, consider the tedious process of crafting an appealing cover letter. A search engine can guide you to articles outlining the key features of a successful letter.

While a search might list '10 tips for an appealing cover letter,' a Generative AI system can do much more. Given your situation, it can generate a persuasive letter tailored to the job you're applying for. It can also help customize the style and tone of the letter to make it unique and engaging, specifically for you.

The difference is palpable: while traditional search engines take us to the doorstep of knowledge, Generative AI welcomes us into the home of understanding.

They find the information we seek and contextualize, interpret, and apply it according to our unique circumstances.

Generative AI transitions us from mere information seekers to empowered knowledge creators, saving us time, reducing cognitive load, and allowing us to focus on implementing the insights rather than gathering them. This radical shift, this game-changing prospect, has captivated the world's attention.

Visualizing The Unseen: Image Synthesis

Generative AI is not confined to the realm of text alone. In the image domain, it promises to redefine our interaction with visuals, giving form and substance to our thoughts, ideas, and descriptions. It is where AI transforms into a visual alchemist, turning our ideas and thoughts into striking visuals.

Let's explore how the "Text to Image" Generative AI systems might shape our future.

Historians, scholars, and enthusiasts of ancient texts often yearn for visual depictions of the narratives within their pages. Imagine feeding an AI system with a description from a centuries-old document — a grand castle with soaring towers, bustling marketplaces, or intricate attire from a bygone era.

The AI system could generate vivid, accurate visualizations, breathing life into these historical narratives. Such capabilities could deepen our understanding of our past, making history more engaging and immersive.

In the realm of law enforcement, the implications are equally profound. Witness descriptions of potential suspects can be notoriously challenging to visualize accurately. However, AI could fill this gap by generating realistic images based on these descriptions by translating textual descriptions into images. This could greatly accelerate criminal investigations, enhancing the effectiveness of law enforcement agencies.

In the dynamic world of e-commerce, Generative AI could democratize high-quality product visualization. Small businesses and emerging brands may lack the resources for professional photoshoots.

AI can create stunning visual representations of products based on textual descriptions, allowing these businesses to showcase their offerings in the best possible light. This could level the playing field for smaller players, boosting their visibility and competitiveness.

The creative industries, too, are poised for a paradigm shift. Artists and graphic designers can use AI to explore and manifest their creativity. By inputting a simple phrase or detailed description, they can create visuals that otherwise might have been impossible or extremely time-consuming. AI, in this sense, becomes a collaborator, aiding artists in realizing their unique visions.

These AI-generated images could be a starting point for their creations, stimulating their imagination and accelerating the ideation process. These AI systems can also take "original images or concepts" from artists and render them in myriads of ways based on textual directions.

The magic of AI doesn't stop here. The film industry could be transformed through the application of Generative AI. Scenes of movies could be initially created and visualized through textual interaction with AI, providing a visual storyboard to guide filmmakers. This could streamline the pre-production process and offer filmmakers a new way to visualize and plan their narratives. Generative AI could unleash a wave of innovation in fashion and interior design. Designers could describe a potential design in words – a dress with an asymmetrical hemline and floral embroidery or a living room with art deco styling – and the AI could generate images based on these descriptions. This could allow designers to rapidly iterate on their ideas and visualize new designs without sketching or modeling them manually.

The real estate and architecture sectors could also harness the power of Generative AI. Imagine property developers being able to feed an AI system with a description of a planned development – the number of buildings, their style, the layout of the gardens, and so on – and have the system produce a detailed, realistic visualization of how the finished development would look. This could help developers and architects present their visions to clients and stakeholders more engagingly.

In the world of gaming, Generative AI could become an integral part of the game development process. Developers could describe scenes, characters, or environments, and the AI could generate images that could then be refined and incorporated into games. This could lead to a new era of personalized gaming, where players can customize their gaming experience by describing their ideal characters or environments and bringing them to life through AI.

In advertising, Generative AI could be used to generate highly customized and personalized ad visuals. Marketers could input descriptions of their target audience, the product, and the desired emotional response, and the AI could generate visuals that hit the mark. Turning text into images could become a powerful tool for visualizing the unseen, expressing the unspoken, and giving life to our wildest ideas.

Talking to Spreadsheets: Generative AI and Data Analysis

Numbers, figures, tables, spreadsheets – these are the lifelines of any organization. They are the undercurrents driving decisions, shaping strategies, and influencing outcomes. Yet, the depth of this data sea can often be overwhelming, even for the seasoned sailor. Enter Generative AI - a powerful ally in navigating the vast oceans of data.

Imagine a world where spreadsheets do not just passively display numbers. Instead, they actively engage with you, translating your spoken or written thoughts into analytical models. In this world, the AI-enabled spreadsheet evolves into an insightful partner that understands your needs, anticipates your questions, and delivers concise, relevant insights.

Take, for example, a CFO steering the financial ship of a large multinational corporation. There is a sea of data to navigate – balance sheets, income statements, cash flow statements, sales reports, and more. With Generative AI, the CFO doesn't just see the data. Instead, they converse with it. Suppose the CFO muses, "I need to understand the effect of a 5% increase in marketing budget on our bottom line over the next quarter." The AI assistant, understanding the context and dependencies, can generate a scenario that forecasts the implications. Or perhaps they ask, "Can you model the financial impact if we shift production from our Chinese factory to the Brazilian one, factoring in the new tax regulations?" The AI generates a comprehensive financial model highlighting potential gains or losses in seconds.

Furthermore, the system can respond to dynamic analysis requests. Instead of manually sifting through spreadsheets or navigating pre-built dashboards, AI could generate a living, breathing analysis responding to user queries. Executives could simply ask the AI questions about their data. "What's our expected financial runway?" "How do our profit margins compare across different product lines?" "Which department is incurring the highest expenses?" "What are our top 5 expense heads for the last quarter, and how do they compare to the same quarter last year?". The AI system can promptly pull up this data, comparing and analyzing the information to give meaningful insights. Imagine the increase in effectiveness when all these can happen during a meeting or a strategic discussion.

The Art of Code Conversations

Software development is an intricate blend of logic, creativity, problem-solving, and relentless attention to detail. For developers, it's like creating an intricate tapestry of code, interweaving different programming languages, tools, and frameworks. Generative AI has the potential to transform this landscape and enrich it significantly.

Imagine this: a developer is working on a complex machine-learning application. They have a rough idea about the overall structure, but the implementation details are hazy. Traditionally, this would involve several hours, even days, of meticulous coding, debugging, and refinement.

However, with a Generative AI tool, the developer simply explains the design and functionality in everyday language. The AI assistant then promptly generates potential codes, algorithms, or design patterns that the developer can tweak and refine. In this process, let's not forget that the first part of figuring out the design and functionality is already a big challenge for developers and product managers.

Let's say a developer is working on a challenging machine-learning task. They can instruct the AI to "Generate a Python function to classify the dataset into ten labelled classes using a Random Forest algorithm ." Understanding the requirements, the AI assistant could produce a skeleton code that the developer can further refine.

Generative AI can serve as an "on-demand pair programmer," capable of offering suggestions, spotting potential bugs, and proposing optimization strategies. Searching for a missing semicolon or diagnosing a hard-to-spot memory leak becomes significantly less burdensome when you have an AI assistant that can instantly scan millions of lines of code and detect anomalies.

Imagine an AI-powered co-programmer sitting by your side, ready to translate your high-level thoughts into precise code snippets. The co-programmer understands various languages, libraries, and frameworks, making the developer's job significantly easier. An AI assistant can become a 'junior programmer,' collaborating with the developer in real-time, suggesting potential solutions, and helping debug complex code snippets. It becomes more than a sophisticated "auto-complete."

Moreover, it can significantly expedite the debugging process. Often, developers spend countless hours chasing elusive bugs. In such scenarios, a Generative AI could be a game-changer.

Describe the bug to the **AI:** "The application crashes whenever the user tries to upload an image larger than 5 MB." The AI assistant could then suggest potential areas in the code that may be causing the issue or even provide fixes to the problem.

Even more fascinating is the idea of "speaking" a website into existence. Imagine a business owner who wants to develop a website without programming skills. With Generative AI, they can describe their vision for the website"—I want a minimalist-style homepage with a rotating banner showcasing our latest products, a navigation bar at the top, and a customer testimonial section towards the bottom." The AI could then generate a working mock-up based on this description, providing multiple design options for selection.

An example of this could be, "I want to build an e-commerce website for a local bakery specializing in artisan bread. The site should have a rustic feel, and the user interface should be intuitive." Understanding your needs, the AI generates multiple design options for you, complete with color schemes, typography, and user flow.

Moreover, AI could potentially revolutionize code maintenance and understanding of legacy systems. When new developers join a project, they often struggle to understand the existing codebase. Here, the AI could analyze the code and provide a comprehensive overview of the software's structure, functionality, and dependencies. It could identify areas where the code could be optimized or suggest potential improvements, making the onboarding process smoother and more efficient.

Generative AI can transform software development from a laborious, linear process to an interactive, dynamic conversation. As AI better understands context, semantics, and user intent, it could potentially handle more complex programming tasks, further extending its role as an indispensable aide to developers.

Demystifying Genomics

Genomics is often described as the language of life, each gene and protein a word in a vast, complex, and as yet partially deciphered book of life. As Generative AI makes headway into genomics and biotechnology, we are unlocking a powerful tool that can read, understand, and even write in this language.

The ability to converse with an AI system in the realm of genomics is akin to having a powerful computational microscope. Imagine querying the AI, "What protein structures could potentially bind to this receptor for therapeutic purposes?" The AI could generate protein structures by understanding protein folding principles and available data. The implications for drug discovery are profound. An iterative, interactive dialogue with the AI could identify a lead candidate faster than traditional methods, saving years of research and billions of dollars. Researchers could ask an AI system to "Predict the folding patterns for this amino acid sequence," and the system could generate a 3D model of the protein's structure. This model could then be used to understand the protein's function or how it might interact with potential drugs. The impact of this application was highlighted in 2020 when Google's DeepMind used their AI model, AlphaFold, to predict protein structures, heralding a new era in biology and medicine.

Consider the possibilities in gene synthesis. A researcher could ask the AI, "Can you generate possible gene sequences that might produce a protein with these characteristics, drawing insights from this specific species?" AI's ability to crunch vast volumes of data and draw connections between disparate information points could open new avenues in synthetic biology. Going beyond, Generative AI can aid in developing vaccines.

The researcher could query, "Can you predict the possible variations of this virus and the potential impact on the vaccine's efficacy?" Such insights could help anticipate and prepare for viral mutations, accelerating the development of future vaccines.

While the AI systems will not give exact answers and groundbreaking discoveries, they could hasten the process for human geniuses, assisting them with the power of large-scale semantic understanding of this complex domain.

The power lies not only in the algorithms that generate different genomic structures or protein folding. While these algorithms have been constantly becoming sophisticated, their added ability to interact with them through "natural language" will alter the landscape forever.

AI Applications in Everyday Life

Coming down from the realm of possibilities, Let's get a hands-on feeling of what these developments in AI will mean in our everyday lives. Let's look at how mainstream apps and digital services we use daily will soon be transformed. Let's glance at the crystal ball and consider the evolution of social media with generative AI.

Imagine this: you're posting an Instagram story of your beach-side sunset. Instead of scouring your music library or Instagram's suggested music, ask your AI, "Could you generate a chill, lo-fi track that would suit a beach sunset vibe?"

The AI responds by crafting an original melody that perfectly complements the tranquil visuals. Not only does this enhance your story, but it also provides a unique auditory experience for your followers.

Next, you could ask your AI assistant instead of wracking your brains to develop catchy, trending hashtags. Based on the context of your post, the AI could generate trending and relevant hashtags that increase your post's visibility. No more manual searching; just a natural, human-like conversation to get you the desired results.

Now, let's bring this AI magic to the dating world. Tinder and other dating apps could be transformed drastically. Writing engaging conversation starters or creating a compelling profile can be daunting tasks. With AI, you could ask, "How can I make my profile more interesting?" or "What's a witty opening line for this match?" Imagine being able to tell your AI, "I want to start a conversation about their interest in mural painting in a witty and respectful way.". The AI could then generate personalized suggestions based on your profile, preferences, and your match's interests. This could help break the ice, providing a more natural and comfortable way to initiate interactions.

Creating an appealing Tinder profile could also become effortless. Instead of hiring a professional photographer, you could tell your AI, "I want a profile picture that reflects my love for adventure and travel."

The AI could then generate an image of you skydiving, hiking, or exploring exotic places, all based on your existing pictures and the data it has on such activities. We can also imagine interactions like: "Can you suggest a backdrop for my picture that complements my outfit?" or "Can you modify the lighting to highlight my features better?" The AI, understanding your requirements and the aesthetics of a good picture, could generate multiple options for you to choose from.

Generative AI is set to blend seamlessly into everyday digital interactions, augmenting our experiences and bringing a sense of ease and novelty. These scenarios are just the tip of the iceberg.

Revolutionizing Entertainment: Generative AI in Streaming

Imagine a Friday evening after a long work week. You're exhausted and only want to unwind with a good movie or TV show, but you're unsure what to watch. Navigating endless genres, titles, and episodes feels as taxing as the work you just left behind. Now, picture a different scenario: you simply voice your mood or preferences, and a specially curated list of content, finely tuned to your tastes and current state of mind, appears. That's the power of generative AI in streaming services.

Platforms like Netflix could elevate user experience by leveraging generative AI. Rather than scrolling through rows of categories or getting lost in an algorithm-suggested list of shows, users could engage in a text or voice conversation with the AI. "I'm in the mood for a heartwarming movie," or "I want to watch a thriller that will keep me on the edge of my seat," could become common commands.

Armed with an in-depth understanding of movie scripts, plot arcs, character developments, and cinematographic elements, the AI would not merely suggest random titles in the requested genre.

Instead, it would present a collection of carefully selected movies or shows that resonate with the user's specified mood or interest. It could also introduce users to lesser-known titles or indie productions they might not have discovered otherwise, enriching their cinematic experience and broadening their horizons.

For instance, you could tell the AI, "I'm in the mood for a quirky comedy that features a road trip." Based on this input, the AI could scour the Netflix library and return with a list of titles you might enjoy, such as "Little Miss Sunshine" or "We're The Millers." Moreover, it could even recommend lesser-known titles that might have escaped your attention, expanding your cinematic horizons.

This application of generative AI moves beyond mere recommendation algorithms. It represents a shift from reactive suggestions based on past viewing habits to proactive, sophisticated engagement based on real-time preferences and feelings.

Similar advancements could redefine how we discover music on platforms like Spotify. Today, genre and past listening habits largely determine playlists and song suggestions. Generative AI, however, could transform this into a more engaging, conversational process.

As with movies or TV shows, users could describe their current mood, desired atmosphere, or a type of music they're curious to explore. The AI could generate a playlist that precisely fits the criteria, creating a uniquely personalized auditory experience.

Imagine telling the AI, "I want to hear music that feels like a rainy day in the countryside, or something that evokes the spirit of 60s rock n roll, but with a modern twist." With this detailed, mood-oriented input, the AI could delve into Spotify's vast library and generate a customized playlist featuring songs that perfectly match your unique request.

Instead of being constrained by traditional genre labels or artist tags, music lovers could navigate the vast universe of songs and compositions through intuitive and expressive language. "For example, Play me songs like the wind rustling through autumn leaves" or "I want to hear music that feels like a warm hug on a cold day" could prompt the AI to curate a unique playlist, painting an aural picture that perfectly suits the user's poetic request.

We envision a future where our devices understand not just what we've enjoyed in the past but what we're craving in the present moment, responding with tailored entertainment options that hit the right notes, visually and audibly.

With generative AI, streaming services could transform into interactive, personalized companions, understanding and catering to our mood swings, whims, and fancies like never before. The future of entertainment is not just personalized; it's conversational, intuitive, and in tune with our emotions.

Redefining Correspondence: Generative AI in Emails

Correspondence can often seem like a chore in the rush of our busy lives. Drafting formal emails or penning polite customer support queries requires a certain level of time and focus that we may not always have. Imagine, then, a world where this hassle is eliminated.

Enter the realm of generative AI, where the future of email writing is not only convenient but also elegantly efficient.

Powered by generative AI, email drafting can be significantly streamlined. Rather than spending precious minutes composing and refining a message, users could simply list a few bullet points outlining their main ideas or concerns. The AI could then weave these points into a coherent, well-structured email that perfectly mirrors the user's unique writing style or even improves upon it.

For example, dealing with customer support often entails a back-and-forth exchange of messages that can be frustrating and time-consuming. With generative AI, this process becomes a breeze. You could tell the AI, "I want to return a product because it arrived damaged," or "My Internet connection has been unstable for three days." With this input, the AI could construct a detailed, polite, and assertive message articulating your issue and asking for a solution.

The applications of generative AI in email don't stop at drafting messages. The technology could also generate auto-responses or out-of-office replies that go beyond the standard, impersonal templates we're used to. Instead of a bland "I'm currently out of the office with limited access to email," your AI assistant could craft a personalized, friendly message that better maintains your professional relationships during your absence.

Moreover, generative AI could assist in managing the volume of emails we receive daily. By understanding your preferences and priorities, the AI could highlight the most important messages, suggest quick replies, or draft longer responses for your review.

As we end this illuminating journey, it's clear that the future we've been discussing is neither a far-off dream nor an ephemeral fad. It is a future steeped in reality, shaped by the immense potential of generative AI. The narrative woven across multiple domains is not merely a work of science fiction. Many of these possibilities are already within our reach, while others are on the horizon, destined to be realized in the coming years.

From transforming our daily digital interactions to catalyzing unprecedented advancements in fields as diverse as genomics, finance, and software development, the impact of generative AI is undeniably profound. Our journey through this chapter has not been a voyage through a fantasy dreamland; rather, it has been a vivid exploration of a future closer than we might think. While the scale and scope of the applications we discussed might seem dizzying, it's essential to recognize that many of these innovations are already in motion. We are not merely speculating about a distant reality; we're observing the contours of an impending future taking shape.

In this exciting yet complex landscape, we are not just passive spectators; we are participants, pioneers even in the emergence of a new world. The 'gold rush' days of AI development are upon us; with it, we must navigate the opportunities and challenges. Ethical, societal, practical, and legal considerations must be addressed concurrently as this technology continues to evolve and permeate our lives.

Of course, this transformation won't occur without disruption, and it's natural to contemplate the potential impact on jobs and livelihoods. The reality of technology-induced displacement is a critical issue we will delve deeper into in the upcoming chapters.

The wave of generative AI is undeniably real, and turning a blind eye to its rise is not an option. Rather, we must embrace the opportunities it presents, learn to navigate its complexities, and actively participate in shaping this new era. Hence, this book - and our journey through it - holds profound significance. We're not just learning about generative AI; we're preparing to thrive in a world transformed by it.

4

UNDERSTANDING THE DISRUPTION

The Dawn of AI - A Symphony or A Storm?

In an age where dazzling marketing jargon dances across screens and enthralling visual art captures our collective imagination, a more silent, profound revolution brews. This revolution is not confined to spectacular graphics or ingenious marketing campaigns; it is about redefining our relationship with machines. It's almost poetic: machines, once mere executors of our commands, now stand on the precipice of becoming co-creators, partners in the very essence of human invention and creativity. And at the heart of this seismic shift? Artificial Intelligence.

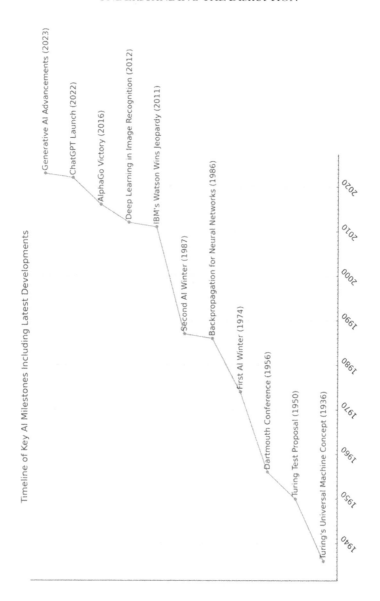

Gone are the days when AI was dismissed as a fleeting trend, a mere blip in the ever-evolving tech panorama. If our journey so far through the realms of AI has taught us anything, this entity is no passing fancy. It has nestled itself firmly into the foundations of our digital lives. AI's omnipresence has been evident as we've voyaged through the last chapters. From the intricate arts of copywriting, the masterpiece paintings in galleries, to the meticulously coded software, the intricacies of medicine, the discernment in law, and even the mysterious fabric of genomics - AI isn't just a bystander. It's a protagonist setting the stage for a future that is as exhilarating and, occasionally, terrifying. As we marvel at AI's majestic vista, we're equally confronted by the shadows it casts—shadows that are often darker, deeper, and more disruptive than we'd ever imagined.

As you sit in your ergonomic chairs, typing away at jobs you consider safely ensconced in the realm of 'human expertise,' there's a seeping notion that perhaps AI has other plans. Plans that might not include your career's longevity. Busy crunching numbers for the next fiscal year, executives are equally anxious, scrambling to retrofit entire enterprises for an AI-augmented future. The 'next big thing' is not merely a blip on an investor's radar; it's a cosmic event they're desperate not to miss. Even the corridors of power are abuzz as politicians and policymakers grapple—often clumsily—with questions with no precedent, template, or easy answers.

Understanding, it's said, is the panacea to fear. The unknown, the unseen, often casts the darkest shadows in our minds. As we delve deeper into this tome, our quest will be just that – a quest for clarity. We'll dissect the staggering impact of AI on our lives and societies, peeling back its many layers and delineating opportunities that will enable us to be masters of this incredible potential and not bystanders trampled by its progress. We'll navigate it's 'what,' unravel it's 'how,' and probe it's 'why.' Our hope? As we turn the last page, we stand ready for a future powered by AI - not cloaked in fear but armed with clarity.

The seed of every technological advancement is sown in the fertile ground of the human mind. It all begins with recognizing a task that's too challenging, time-consuming, or monotonous. As if led by a primal urge to problem-solve, the inventor seeks to create a tool that could make life 'better' in some significant way. It's a spark of inspiration that compels us to say, "There has to be a better way." And in that instant, we set ourselves on a path of creation, motivated by the pursuit of improvement, be it for efficiency, speed, or simplicity.

Consider the humble wheel. The first potter who crafted it might not have fully grasped the extent of the revolution they had set into motion. The wheel started as a simple device to move pots around more effortlessly, but quickly, its utility became evident. Over time, it evolved into chariots and carts, forever changing how goods and people moved across distances. This transformative tool paved the way for developments in countless other domains, heralding an era of transportation that revolutionized societies and economies.

The genesis of a single idea, the conception of a core invention, and its birth are subtle yet profound events, a creative spark ignited in the solitude of a single mind. However, this invention will not remain confined to these boundaries once its utility is realized. It starts spreading like wildfire. It's used, shared, tested, and refined. It's adapted and adopted for tasks of a larger scale and in domains beyond its original design. Consider the personal computer. Initially, computers were massive machines confined to government agencies and large corporations. They were expensive, difficult to use, and not something the average person could own or operate. Then came the idea of a personal computer for everyone. This transformative tool changed the way we work, play, and communicate. It started as a tool for computing tasks but quickly became so much more. Today, it's a multimedia device, an entertainment center, a communication hub, a library, a shopping portal, and so much more. The growth trajectory for the potter's wheel, the printing press, and the personal computer's path from conception to ubiquity followed the same pattern. An idea became a tool that found a purpose and demand, leading to refinement, mass production, and widespread adoption. This was the case for the wheel, the steam engine, the Internet, and countless other inventions that have shaped our world.

As the value of an invention becomes increasingly apparent, its appeal grows, sparking a ripple effect that accelerates its evolution. The demand for it swells in the marketplace as more people are willing to pay for its benefits. This growth, in turn, attracts more creators and companies who invest time, money, and energy into making it on a larger scale. This mass production decreases the cost, making the technology accessible to even more people and further amplifying its utility and ubiquity.

This cyclical, self-perpetuating process fuels the advancement of tools and technology, a process as old as human civilization. At the heart of it all is a rather understated yet critical principle - the relationship between market demand and utility. Technology's value and the balance it strikes between its cost and the benefits it offers ultimately determine its fate.

We've seen it repeatedly - technologies that serve a real need to survive and thrive. Regardless of how cool or innovative they may seem, those who can't meet this criterion slowly fade into obscurity. The marketplace of ideas is ruthless and pragmatic. Only the tools and technologies that deliver genuine, tangible value endure the test of time.

In this ever-spinning wheel of innovation, driven by necessity, refined by use, propelled by market demand, and filtered by the unyielding principle of utility, we can glimpse the invisible hand that has shaped our past and is ceaselessly carving our future. This lens shows a series of inventions and an ongoing dialogue between humans and technology. This conversation redefines our world and our place within it.

The unvarnished truth is that economic incentive has always been. It will likely always be, the primary engine driving technological adoption, shaping the destinies of nations, corporations, and individuals alike. When a technology—like Artificial Intelligence—matures to the point where its economic advantages become too colossal to ignore, it's not merely an option for enterprises to adopt it; it becomes an existential imperative. As Artificial Intelligence inches closer to its full-blown maturity, replete with unprecedented economic incentives, small and large businesses will find themselves at a consequential crossroads: to embrace AI-driven metamorphosis or risk obsolescence.

Yet, it's not merely a matter of choice; it's the inexorable pull of Darwinian economic survival. Just as it's nearly inconceivable today to picture a competitive organization operating sans electricity or the internet, we're swiftly approaching a juncture where sidestepping AI will be equivalent to signing an enterprise's extinction certificate. As this wave grows more formidable, its repercussions on the workforce will be undeniable. As we know them today, jobs are poised for a paradigm shift. For the workforce, this metamorphosis will herald an era of displacement as positions once 'irreplaceable' become automated, requiring us to confront a new landscape of professional viability.

But let's set aside the overused and vaguely apocalyptic phrases like "AI will rule the world," or "prepare for a Terminator takeover." Instead, let's sharpen our focus and delve into the specifics. What kinds of jobs are at risk of automation through AI? Why will these particular roles become redundant? The answers to these questions will illuminate the challenges ahead and guide us in crafting strategies for pivoting to remain indispensable.

Transformative And Frontier Technologies

Every wave of technological advancement brings with it a tide of disruption. As new tools mature and become widely adopted, the traditional methods they replace are often rendered obsolete. In the world of technological advancement, this constant oscillation between anxiety and anticipation is often based on the technology being introduced. To understand the true impact of these innovations on job markets and human potential, it's essential to classify technology into broader categories. Here, we'll delve into two primary classes: Transformative Technologies and Frontier Technologies.

Our innate desire to improve, evolve, and amplify is the core of human progress. Transformative technologies tap into this very desire, serving as tools that enhance our natural abilities. Imagine a singer wishing to reach a larger audience. While their voice may be powerful, the aid of a microphone amplifies that power, allowing them to reach thousands in an arena. In essence, transformative technologies are like that microphone, magnifying what is inherently human. Their primary purpose is amplification – taking an inherent human capability and elevating it to new heights. They're not about reinventing the wheel; they're about making that wheel spin faster, move more weight, and tread paths previously thought unreachable.

Consider the evolution of communication. From the primitive drums and smoke signals to the telegraphs and then to the modern-day smartphones, each leap was transformative. While the fundamental human desire was to communicate across distances, transformative technology made it quicker, more precise, and more efficient to transfer large volumes of information across the globe within seconds.

However, there's a critical facet to these transformative technologies that we cannot ignore: their potential to replace human roles. Take the realm of agriculture, once the backbone of economies and societies. The advent of agricultural machines streamlined processes that once required entire communities. With the birth of the combine harvester, fields that took weeks to reap could be harvested in a day or two. The efficiency was undeniable, but so was the sharp decline in the workforce needed. In the manufacturing sector, the story was much the same. The onset of the Industrial Revolution saw machine looms replacing manual ones, rendering countless textile workers redundant. When they entered the corporate and industrial world, computers took over roles that were once exclusively human. With their precision and efficiency, industrial robots further reduced the need for human intervention in assembly lines and quality control.

The certainty of these technologies lies in their inherent benefits: the ability to perform tasks on an unparalleled scale, with consistency, precision, and, often, significant economic efficiency. Industries naturally gravitate towards these innovations in their quest for progress and productivity. The scale tip favors transformative technologies because they amplify human capabilities, so the original, unaided human skill seems almost trivial. Machines and algorithms, once merely assisting hands, become the primary workforce. They operate ceaselessly, without the constraints of fatigue or emotions, making them economical business choices. As a result, the number of human jobs these technologies replace invariably exceeds the new roles they create. The math is simple: when one machine can perform the task of a hundred people, even if its integration and development creates a few specialized jobs, there's a net loss in employment.

If transformative technologies serve as mirrors, reflecting and enhancing human capabilities, frontier technologies are the telescopes, revealing the vast unknowns and charting a course through uncharted territories. These are the true pioneers of human ingenuity, breaking barriers and allowing us to venture into realms previously thought to be the stuff of myths or mere dreams.

Frontier technologies don't just iterate on what we know; they radically redefine what we know is possible. These technological marvels pull back the veil on new dimensions of knowledge and capability, birthing industries and job roles that previous generations could have only fantasized about.

Take metallurgy, for instance. Before its advent, humanity was confined to the constraints of naturally available materials. With the dawn of metallurgy, we didn't just improve our tools; we reimagined them. Suddenly, human civilization had the means to craft tools, weapons, and structures of unprecedented durability and function. More than just a craft, metallurgy became a thriving industry, giving rise to countless professions – from miners to blacksmiths to metal artists.

Nuclear technology, another marvel of the frontier category, did more than provide an alternative power source. It reshaped our understanding of energy and matter at a fundamental level. As we harnessed the power of the atom, industries dedicated to energy production, medical applications, and research blossomed. Nuclear physicists, reactor operators, and radiation therapists emerged as crucial professions, catering to needs humanity wasn't even aware of a century ago. The wonders of space technology propelled our

aspirations literally to new heights. A domain once reserved for astronomers and dreamers became a bustling industry. Satellite technicians, aerospace engineers, and even space tourism became a reality. Space exploration brought advancements in telecommunications and meteorology and opened up the tantalizing possibility of extraterrestrial colonization and exploration.

Biotechnology has unlocked new doors in our understanding of life itself. We transitioned from passive observers of nature's miracles to active participants in its design. Nanotechnology, operating at scales previously invisible to human eyes, has ushered in a revolution in material science, electronics, and medicine. As we began manipulating matter at the atomic level, we unlocked potential applications that seemed almost magical.

Frontier technologies are not just advancements; they're evolutions. They don't replace; they introduce. Unlike their transformative counterparts, which might reduce the need for human labor in existing domains, frontier technologies birth entirely new domains teeming with opportunities. They remind us that the story of human progress is about refining our present and redefining our future.

Every so often, the wheel of progress turns not by a degree but by a revolution. A technology emerges that doesn't just fit neatly into a predefined category but expands the very horizons of those categories. These revolutionary technologies stand at the crossroads of transformative and frontier innovations, becoming keystones of civilizations in profound and all-encompassing ways.

Electricity is the quintessential example of such a technological marvel. To deem it merely transformative would be to undersell its omnipresent influence on daily life. It didn't just amplify or replace an existing human capability; it fundamentally rewired our societal structures. On the other hand, calling it solely a frontier technology doesn't quite capture its essence either. While it did introduce us to entirely new possibilities, its magic lies in its ubiquity, touching every facet of our existence.

Imagine the world in the early 20th century. Darkness was beaten back by flickering candles or gas lamps. Communication was largely manual and limited to the speed of a horse or a steamship. Industries relied on steam, water, or manual labor. Fast-forward to today and our nights are illuminated, our homes powered, and our industries turbocharged, all by this invisible force. Such is the pervasive power of revolutionary technologies; they transform the very fabric of societies in the blink of a historical eye.

Another titan in this category is digital computing. At its inception, computing aimed to augment human calculation abilities – a transformative goal. Yet, its evolution has transcended this initial scope, making it a cornerstone of frontier ventures. It's not just about faster calculations anymore; it's about recreating and redefining reality. The realm of zeros and ones serves as the foundation upon which the dreams of virtual reality, advanced AI, and global connectivity are built. Simulations that mirror real-world complexities, virtual landscapes that defy physical limitations, and instantaneous global communications are no longer feats of fantasy but everyday occurrences.

Moreover, digital computing has seamlessly fused with other frontier technologies, catalyzing their advancements. The meticulous genetic sequences in biotechnology, the accurate star maps in space tech, and the atomic precision in nanotech – all owe a nod to the computing prowess that underlies them. Its capacity to model, predict, and design has enabled humans to venture confidently into the unknown, armed with data and insights.

What makes these technologies – electricity and digital computing – truly extraordinary is their unparalleled versatility and pervasiveness. They become so integral that imagining a world without them becomes challenging and almost inconceivable. In their wake, they don't merely leave behind transformed industries or new frontiers; they redefine the concept of possibility. Their legacy is not one of displacement or creation but of boundless expansion.

The AI Epoch – Skill Scope Canvas

As we stand at the cusp of a new technological age, the conversation invariably gravitates toward Artificial Intelligence, specifically generative AI. But what is the true essence of AI within our dichotomy of transformative and frontier technologies? How does it fit into our narrative of human progress, and where will it steer our collective journey?

At its core, AI, particularly generative AI, is an unparalleled maestro of pattern recognition. It dives deep into the vast oceans of human data, discerning intricate patterns and then weaving those findings into new, unseen tapestries of information. On the surface, this makes AI the epitome of a transformative technology. It acts as a colossal cognitive amplifier, taking the inherent human ability to identify patterns and elevating it to scales previously thought impossible. It's as if our brain has been given a supercharger with all its complexities and nuances.

One might argue that AI doesn't unlock new frontiers of human understanding in the traditional sense. After all, it thrives on data created by humans. However, merely labeling AI as transformative would be an injustice to its potential. Consider this: the very nature of AI, its ability to combine, recombine, and generate entirely new patterns and solutions, could lead to insights previously inaccessible to the human mind. The true genius of AI, especially in conjunction with digital computing, is its foundational capacity. Like the electrified grids that power cities or the binary code that underpins our digital universe, AI promises to be the substrate upon which future technological marvels will rise. Today's applications – in healthcare diagnostics, financial predictions, or creative arts – are just the preliminary brush strokes on a canvas that will host a masterpiece of human-AI collaboration.

While AI will soon see massive transformative effects, leading to massive job loss, evident in the efficiency and scale it brings to tasks, its frontier potentials are vast and largely uncharted. Its true potential lies in the problems it solves today and the doors it will unlock tomorrow. The evolution of AI could lead to the creation of tools and technologies we have yet to conceive. It is akin to the wheel inventor not being able to foresee the advent of motorized vehicles, let alone autonomous ones.

To better understand the nature of impending job loss due to AI technologies, let's paint a comprehensive picture of today's professional landscape. We will map various jobs into a Skill Scope Canvas (SSC), which maps the progression of jobs from the purely physical to the predominantly mental while simultaneously layering in the dimension of task sophistication.

SKILL SCOPE CANVAS

Increasing sophistication level (High → Low)

Highly Physical → Highly Mental

	Highly Physical →					→ Highly Mental
High	Athlete Astronaut Military Pilot	Artist Cheft Archeologist	Surgeon Architect Event Manager	Entrepreneurs Inventors Coach	Judge Politician Policy Makers	Executives Investor Economist	Scientists Philosopher Priest
	Firefighter Soldier Goldsmith	Photographer Physio Carpenter	Farmer Travel Guide Police	Detective Engineers Supply chain	HR Manager Author Anthropologist	Journalist Marketer PR	Software Architects AI Engineers Therapist
	Tailor Factory Worker Janitor	Gardner Painter Mason	Office Clerk Retail Worker Delivery Agent	Gen Physician Teacher Junior Designers	Accountant Lawyer Insurance Agents	Copywriters Graphic Designer Market research	Data Science Stock Traders Junior Programmer
Low	Security Guard Warehouse worker	Bank Teller Fastfood cook	Driver Waiter	Receptionist Office Intern	Secretary Data entry Agents	Proof readers Financial Analyst	Web Developer Call Centre agent

Increasing sophistication level

* This is a representational list, not an exhaustive one

The left end of this canvas symbolizes highly physical tasks. As we move rightwards, we see a shift towards increasingly mental jobs. Now, visualize a vertical spectrum that traverses this canvas from bottom to top. This is our sophistication gradient. Jobs at the bottom of the canvas are simpler, requiring lower skill levels, while the ones at the top demand high training, expertise, and sophistication.

As we populate this canvas with professions, we embark on a journey through the dimensions of physicality, intellect, and sophistication. In the lower-left area of our canvas, we encounter professions like construction workers, assembly-line workers, or miners. These jobs require significant physical labor but relatively less sophisticated skills. As we move upwards along the left-hand side, we find professions that demand both physical skill and a higher degree of training or expertise. This includes athletes, surgeons, or craftsmen, where physical labor is combined with a significant degree of training, skill, and often, creativity.

Crossing over to the right-hand side of the canvas, we venture into the realm of mental labor. In the lower right region, we find jobs that require basic cognitive skills but lower levels of sophistication, such as data entry clerks or customer service representatives. As we ascend the canvas along the right side, we witness increasing sophistication in mental tasks. Professions like software engineers, research scientists, economists, or corporate strategists in this zone demand a mental orientation and a high degree of intellectual rigor, training, and creativity.

> **We have been phasing out jobs on the left side, mainly at the bottom. Many of the past jobs are not even relevant anymore to be here. At the same time, we are continuously creating new jobs on the right side, all across the sophistication spectrum. A "data labeler" job didn't even exist ten years ago.**

Returning to the beginning of the 20th century, we notice that most jobs gravitated toward the left side of our Skill Scope Canvas. Heavily, physical occupations dominated the landscape of labor. From the sweat-drenched miner chiseling coal deep under the earth, the weaver industriously transforming raw cotton into wearable fabric, to the tireless farmer, his life rhythmically tuned to the seasons and the needs of his crops, the norm was work that predominantly engaged the body. These jobs harnessed the elemental human strength to shape the environment, cultivate the earth, construct shelter, and produce goods.

During the first half of the 20th century, the world was a theatre of enormous geopolitical upheaval. The gruesome destruction of the two World Wars, coupled with the political tension of the Cold War that followed, significantly accelerated technological development. It was an era that necessitated and spawned radical technological advancements, initially driven by warfare and later appropriated for civilian use. Whether it was the radar, nuclear technology, or the rocketry that would eventually transport humans to the moon, the crucible of conflict ignited rapid technological progress.

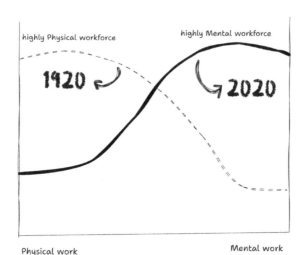

Concurrently, the advent of assembly-line manufacturing, introduced by Henry Ford and his Model T, triggered a dramatic shift in work. As machines began to take over the heavy lifting, manual labor was replaced by semi-skilled tasks that required more mental engagement, albeit at a basic level.

Workers on the assembly line didn't need the strength of a blacksmith. Still, they needed the mental acuity to monitor the process, operate machinery, and maintain quality. We thus see the first significant migration on our SSC, from the extreme left towards the center.

This shift, like jobs, gathered momentum with the dawn of the Information Age in the second half of the 20th century. The invention of the transistor, the mainframe computer, and the internet marked a transformative era in human history.

Information became the new currency, and the ability to process and manipulate this information became a coveted skill. Jobs started to demand more mental exertion, focusing less on physical prowess. As a result, we see a further drift towards the right, especially in the developed world.

Today, many jobs exist on the right-hand side of the spectrum, far removed from their physically-intensive origins. From software developers who design sophisticated algorithms and financial analysts who navigate the complex world of stocks and bonds to designers who create virtual realities, the nature of work has undergone a seismic shift.

In the 21st century, the workplace is almost unrecognizable from its early 20th-century counterpart. The hammer, pickaxe, and plow of yesteryears have given way to the keyboard, the stylus, and the software code. The symphony of work now plays out more in the mind and less in the muscle.

What Jobs Are Disrupted and Why?

Automation and job transition to machines start from the bottom of the sophistication domain. In the physical realm, automation has been happening for centuries. We've replaced the work of human muscles with machines, from the simple lever and pulley to sophisticated robotic arms in modern manufacturing facilities. Indeed, it is often the vision of robots—tangible, physical entities—that we conjure up when we think of automation. We imagine a future where robots clean our houses, cook our meals, and perform other mundane tasks—images propagated by science fiction and Hollywood.

As the human skills from right to left of the scale move from "highly mental" to "highly physical," the machines that will perform these tasks will also need to embody a continuum from "mental to physical" capabilities. These will vary from "invisible software bots in the cloud" to "smart agents with speakers, microphones, and display connected to cloud bots" to "physical robots equipped with advanced sensors, niche actuators, massive batteries, and computing."

Looking through the lens of automation, it's fascinating to see how the nature of the machines evolves along this continuum. At one extreme, on the left side, the domain of heavy physical tasks is largely dominated by hardware machines. These are the mechanical arms in manufacturing plants, the conveyor belts in warehouses, and the tractors in fields, all replacing heavy physical labor with machinery.

In the middle of this spectrum, we find a blend of physical and cognitive tasks performed by increasingly sophisticated robots. These robots aren't just hardware or software but a seamless integration of both, equipped with a body to interact with the physical world and a brain (in the form of algorithms) to process information and make decisions. This category includes autonomous vehicles, drone delivery systems, surgical robots, and more. Moving towards the right, we shift from 'hardware' to 'software,' with cognitive tasks being taken over by algorithms running on computers and servers. This is the realm of advanced data analysis, strategic decision-making tools, and complex system management, where the 'muscle' of technology isn't a mechanical arm or wheel but lines of code and computational power.

While these physical robots are becoming more sophisticated and capable, the transformation happening in the mental domain is arguably more significant and far-reaching. This type of automation doesn't look like a robot—it doesn't have a physical form. Instead, it resides in the cloud, crunching numbers, analyzing data, generating reports, and making decisions. It's been called "white-collar automation" or "knowledge work automation," as it's replacing or augmenting mental labor that was previously thought to be the exclusive domain of humans.

Not so long ago, about a decade ago, tasks that required higher cognitive effort were deemed exclusively human. The 'smart' software of the time was limited to well-defined, strictly controlled environments. Tasks like facial recognition or speech generation were in their infancy, requiring a great deal of intricate programming and extensive computational resources, and even then, they lacked reliability and general applicability.

Fast forward to today, and we're witnessing an AI revolution fundamentally reshaping this landscape. A decade ago, pervasive internet connectivity, cloud computing power, abundant data, and sophisticated machine learning algorithms resulted in unfathomable software capabilities. Machine learning models will soon translate languages with human-like fluency, diagnose diseases from medical images with high accuracy, and drive cars through bustling city streets - all tasks that once required substantial cognitive effort.

This dramatic acceleration in AI capabilities distorts the traditional notions of automation and its trajectory. Throughout the 20th century, technological progress primarily focused on augmenting and automating physical labor. Electromechanical machines replaced manual workers in factories, digital computers automated complex calculations, and robotics started making inroads into tasks requiring a blend of physical and cognitive effort. While these advances have been truly transformative, innovation in these areas is plateauing.

On the other hand, AI has entered a phase of exponential growth and value addition, particularly in tasks demanding a higher level of cognitive effort. In the Physical-Mental spectrum we've been discussing, automation is shifting focus toward the right side, where cognitive effort outweighs physical labor.

This shift represents an interesting, counterintuitive trend. While popular imagination is still preoccupied with the idea of robots replacing physical labor, the true frontier of automation has quietly moved into the realm of cognitive work. In the real-world scenario, an example would be that in a "Hospital" the "highly mental load" based roles like "receptionists, mid-level managers, financial and communication" are bound to be replaced much faster than the "highly physical load" based roles like "janitors, nurses, logistics, etc." This phenomenon goes completely against the historical trajectory of job loss from the physical domain.

SKILL SCOPE CANVAS

Increasing sophistication level →

Highly Physical ←→ Highly Mental

Low ←→ High (Increasing sophistication level)

JOB LOSS TO ROBOTS (Very late in the future)

JOB LOSS TO SOFTWARE AI (Very near future)

* This is a representational list, not an exhaustive one

Athlete, Astronaut, Military Pilot	Artist, Chef, Archeologist	Surgeon, Architect, Event Manager	Entrepreneurs, Inventors, Coach	Judge, Politician, Policy Makers	Executives, Investor, Economist	Scientists, Philosopher, Priest
Firefighter, Soldier, Goldsmith	Photographer, Physio, Carpenter	Farmer, Travel Guide, Police	Detective, Engineers, Supply chain	HR Manager, Author, Anthropologist	Journalist, Marketer, PR	Software Architects, AI Engineers, Therapist
Tailor, Factory Worker, Janitor	Gardner, Painter, Mason	Office Clerk, Retail Worker, Delivery Agent	Gen Physician, Teacher, Junior Designers	Accountant, Lawyer, Insurance Agents	Copywriters, Graphic Designer, Market research	Data Science, Stock Traders, Junior Programmer
Security Guard, Warehouse worker	Bank Teller, Fastfood cook	Driver, Waiter	Receptionist, Office Intern	Secretary, Data entry Agents	Proof readers, Financial Analyst	Web Developer, Call Centre agent

As we delve into the factors behind the accelerated development of AI, we cannot overlook the profound influence of digitization. In today's world, the vast majority of mental work, from drafting documents to coding software, has moved into the digital realm. We work in this boundless digital net that interlinks keyboards and screens, steadily assimilating the cloud.

In the not-so-distant past, much of our mental work was done analogously. Documents were handwritten or typed on paper, software was coded through punch cards, and audiovisual content was recorded and processed using magnetic tapes and physical film. However, this landscape has changed drastically with the advent of digital technology. More importantly, this digital work is increasingly being conducted in the cloud, further consolidating the shift towards a digital, interconnected workspace. This proliferation of digital work brings with it an invaluable advantage: data. We generate a colossal data trove as we communicate, make decisions, or even express ourselves artistically in the digital sphere. Each digital document, code commit, video clip, music composition, tweet or Instagram post, public discussion, blog, documentation, meeting minute, presentation, spreadsheet, etc., provides valuable training data for machine learning algorithms. This data is, in essence, the raw material that powers AI algorithms. It provides them with experiences from which they can learn and improve, mimicking the process humans undergo throughout their lives.

As we produce more and more data, the machine learning algorithms that underpin AI technology become ever more adept at deciphering language, processing visuals, recognizing patterns, making decisions, and even emulating creativity. They are, in effect, learning to 'understand' the complexities of our cognitive world, gaining insights into human thought processes and behavior.

An equally significant consequence of digitization is that it transforms the medium of interaction into a unidimensional one. Human-machine interactions are now primarily confined to screens, microphones, and speakers. This unidimensionality simplifies training AI models, as they only need to learn to operate within this constrained environment. For instance, a text-generating AI doesn't need to understand the physical nuances of handwriting—it simply needs to generate sequences of characters displayed on a screen. This simplicity is a marked contrast with tasks that involve a high degree of physical effort. These tasks are inherently multidimensional, requiring interaction with the world through various senses like vision, hearing, and touch. Furthermore, they often involve manipulating physical objects in many different situations, which adds a layer of complexity that AI systems are still learning to handle.

This unidimensional nature of digital interaction significantly streamlines the development of AI. The complexities of dealing with the physical world – the necessity for precise motor control, the need to navigate unpredictable environments, and the multitude of sensory inputs to process – are largely absent in the digital domain.

Consequently, AI can concentrate on enhancing its cognitive capabilities, learning to handle tasks that demand higher mental effort with increasing proficiency. Further, as the digitization trend continues to grow, encompassing new areas and industries, AI's 'learning environment' expands in tandem. The range and diversity of data AI can learn from increases, paving the way for further leaps in capability.

In essence, the digitization of mental work has provided a fertile ground for the exponential development of AI. By giving AI access to an ever-expanding data pool and simplifying the interaction medium, digitization is propelling AI's cognitive capabilities to new heights, rewriting the boundaries of what machines can do.

Secondly, as we have discussed earlier, the marketplace has always been the ultimate crucible where the true value of a technology is tested and validated. Across the annals of human history, countless innovative ideas and inventions have emerged, capturing the collective imagination.

However, only a fraction of these innovations have survived the relentless market scrutiny. The determinant of survival isn't simply about being 'cool' or 'innovative'; rather, it's about striking a balance between cost and utility, complexity and tangible value. The tools and technologies that demonstrate this balance are the ones that endure and shape our world.

Consider the idea of a fully functional robot performing tasks around the house, like washing dishes, folding clothes, tidying up rooms, painting walls, or assembling furniture. The concept seems alluring, almost futuristic. But the allure begins to fade when we dissect the economics behind such a concept. The cost of building, maintaining, and using a robot that can perform these tasks effectively and safely is astronomically high. Furthermore, the physical world's inherent variability and unpredictability add complexity to these tasks, amplifying the technical challenges and costs.

For instance, imagine the task of a hotel janitor. To humans, the task is simple and straightforward, requiring minimal training. Now, picture a robot capable of performing these tasks. It would need multiple cameras for vision, several mechanical arms for manipulation, a robust computational unit for processing, and substantial energy sources. Then, factor in the variability of hotel rooms—different layouts, furniture arrangements, lighting conditions—and the task becomes even more complicated.

Cracking these technical hurdles, even if such a robot were created, it would struggle to deliver the same level of performance as a human at a comparable cost.

On the other hand, consider a task in the cognitive domain—say, bookkeeping. While seemingly sophisticated, this task can be automated with relative ease, given the advances in AI's language and reasoning capabilities. The work of a bookkeeper, in contrast to that of a janitor, is fairly consistent across different organizations. Hence, a software solution that automates bookkeeping can readily scale and adapt to different scenarios. More importantly, it can do so without expensive hardware or energy resources.

Let's say an application is identified where a robot's functionality brings high value, and the economics align. The practical realities of scaling that solution come into play. For instance, every robot is a complex assembly of diverse components, each requiring its supply chain, manufacturing processes, quality control mechanisms, and maintenance protocols. It's not just about designing a robot that can perform a task efficiently and safely. It's also about sourcing the right parts, assembling them under stringent quality standards, and maintaining the overall system throughout its lifecycle. Each stage introduces potential points of failure, compromising the solution's reliability. Moreover, these complexities increase exponentially as we scale the solution from a handful of robots to hundreds or thousands.

The contrast with software-based AI solutions is stark. Indeed, a piece of AI software can take years to build. But it can be replicated almost instantaneously and at negligible cost once built. The software can be deployed on a cloud server, instantly accessible to users across the globe. It can scale seamlessly from serving ten users to serving millions. Unlike physical robots, the complexity and cost of software do not increase significantly with scale. The software does not wear out with use, does not need physical maintenance, and does not suffer from physical manufacturing defects. Its reliability is largely a function of its underlying code and algorithms, and any issues can often be fixed remotely with software updates.

This fundamental difference in the economics of automation in physical versus cognitive tasks is tipping the scales and accelerating the development and adoption of AI in the cognitive domain. The simplicity and scalability of digital AI solutions make them significantly more economically viable than their physical counterparts. Moreover, the vast trove of digital data available in the cognitive domain provides a rich resource for training and improving AI models, further accelerating their development. Contrary to our historical trend of technology replacing jobs in the physical domain, we will now see more mental domain jobs being replaced by machines. In contrast, the physical domain jobs will remain relevant for a long time.

While software-based AI solutions are highly attractive and exciting domains due to technical scaleability and economic profitability, we should not forget that we live in a physical world where superhuman tools are needed to support and sustain us. Apart from the traditional industrial robots and recent warehouse robots, multiple groundbreaking innovations in healthcare, agriculture, and space domains are pushing this difficult frontier ahead by tackling problems worth solving. While these may not be as fiscally attractive in the short run compared to enticing AI software, in the long run, these advanced robots operating in the culmination of physical and mental sophistication will be part and parcel of our survival and sustenance.

Due to shifts in the last century, jobs have become ubiquitous in the mental domain, and the massive economic potential of automating them presents an irresistible proposition for companies. AI technology isn't simply about reducing labor costs; it's about enhancing businesses' value propositions in previously unimaginable ways. As such, companies are investing heavily in AI development, not just as a means of cutting costs but as a strategic imperative for staying competitive in the evolving economic landscape.

The interplay between economics, utility, and technological feasibility is shaping the future of AI, pushing us toward a world where cognitive tasks are increasingly automated. In contrast, physical tasks remain the domain of humans—for now. The landscape is set for a major transformation with an abundance of cognitive jobs available for automation and the massive economic benefits at stake. The question is not if AI will disrupt these jobs but when and how it will do so and how we as a society will adapt and respond to this change.

5

THRIVING AMIDST DISRUPTION: WORKFORCE

As the curtain falls on the acts we've just explored, we stand with an undeniable revelation, one that fuses the technical, economic, and historical facets into a singular, incontrovertible truth: Artificial Intelligence is more than just a disruptor; it's a game-changer for white-collar professions as we understand them today.

But before you envision a rogue AI sneaking into your corner office and booting up in your ergonomic chair, take a moment to dispel such dramatic illusions. No, the AI revolution won't unfold in a vacuum; it won't operate in some parallel economic universe. AI isn't an isolated entity operating in some detached digital dimension. At its core, the economy represents a vast web of human interactions, magnified at the scale of nations and across continents. AI is not some parallel universe; it operates firmly within the confines of our economic systems, magnifying efficiencies and boosting profitability. To be candid, AI has no raison d'être without the human-centric incentives driving its deployment and refinement.

What this means is staggering in its simplicity: Your job won't be snatched away by a machine; rather, it may be supplanted by another human being who has mastered the art of harmonizing with this transformative technology. Picture AI as a 'mental power multiplier,' an advanced loom weaving intricate data patterns instead of fabric. Those who adapt and learn to operate this 'loom' will be the ones who not only survive but flourish in the radically restructured workplace of the future. Just as a handful of skilled operators replaced a legion of weavers during the Industrial Revolution, we'll witness a similar distillation of the workforce, but this time in white-collar sectors where data reigns supreme.

So, how can you transition from a spectator to a star player in this evolving narrative? The answer lies not in blind panic or denial but in understanding the contours of AI: its strengths, weaknesses, and core functionalities. Whether you're a seasoned professional or a newcomer to the job market, the forthcoming pages will serve as your compass—guiding you through strategies and insights to survive and thrive in an AI-augmented world. Transform yourself into a position where AI is not your competitor but your ally, not a harbinger of doom but a herald of unprecedented possibilities.

The Cognitive Divide – Human vs Machine Intelligence

As we revel in AI's fascinating possibilities in human-machine partnerships, it's crucial to remain conscious of an essential boundary. Despite AI's strides in understanding and mimicking human language, it's critical to appreciate the chasm separating human cognition and machine thinking. To navigate this exciting new terrain, we need to understand the contours of this distinction and temper our expectations of what machines, even those equipped with the most advanced AI, can truly understand and feel.

When we, as humans, engage with language, we don't merely process information; we experience it. A line of poetry can transport us to sun-drenched beaches or star-studded galaxies, conjuring up vivid imagery and triggering myriad sensations. Words can evoke human emotions, making us roar with laughter or pull at our heartstrings. They can recall a tapestry of memories, spark inspiration, and sometimes provoke anger. Language, for us, is not just a tool of communication. It's a gateway to a rich, multidimensional world of experiences.

When humans encounter spoken or written words, they trigger complex reactions. Language is an evocative medium capable of stirring a kaleidoscope of emotions and awakening memories tucked away in the recesses of our minds. Depending on our personal experiences, the same words can paint unique landscapes in our minds. Words can transport us into an enchanting wonderland or a haunting abyss, occasionally complete with sensory stimulation that transcends the textual realm.

Consider this example. The phrase "A warm summer evening" might evoke tranquility and contentment in you. You might recall the scent of blooming flowers or the symphony of crickets. However, when inputted into an AI system, the phrase does not evoke any sensory memories or emotional responses. Instead, it triggers a complex web of probabilistic calculations based on context and previous activations.

Words inspire and move us, amuse us, and frighten us. They have the power to ignite sparks of genius or fan the flames of fury. We weave them into stories and songs, riddles and rhymes, imbuing them with a life of their own. This is the magic of language in human cognition, a magic that is intensely personal, deeply emotional, and intrinsically human.

As we gaze toward machines, we must remember this magic doesn't translate the same way. The processing of words in AI systems is fundamentally different. Inputs of words trigger sets of mathematical operations, leading to probabilistic activations and outputs based on its "training data." These activations are shaped by context and previous activations but devoid of personal experience or emotion. When a chatbot seems to empathize with your problem or express joy at your success, it's not because it feels your pain or shares your happiness. It has been trained to respond with specific phrases under specific conditions. The sentence "I'm feeling blue" might lead a well-trained AI to recognize that the speaker is expressing sadness but does not 'feel' that sadness. It cannot. The AI's 'understanding' fundamentally lacks the emotional resonance that characterizes human understanding.

To illustrate this further, imagine talking to a chatbot about a personal loss. It might respond with an empathetic-sounding message, perhaps like, "I'm sorry to hear about your loss. It's okay to feel sad." While this response might seem emotionally tuned, it is important to remember that the chatbot does not 'feel' empathy or sorrow. It is simply generating a response based on its programming and training data.

This dichotomy between human cognition and machine thinking is crucial to remember as we tread the path of technological evolution. While we can create machines that understand and generate human language, we cannot imbue them with the ability to experience emotions, feelings, or sensory impressions. They might mimic our expressions and replicate our responses but won't feel our joys or sorrows.

No matter how advanced, AI systems cannot 'feel' in the human sense. They don't get inspired by a stirring speech nor amused by a clever joke. They don't experience a rush of joy, a pang of sorrow, or a frisson of fear. As of today, and possibly forever, they lack the subjective human experiences that give our world its hues and flavors.

Future AI systems will undoubtedly continue to evolve, eventually reaching a level of logical reasoning that mirrors our own. They will sift through vast information repositories, draw intricate connections, and make complex decisions, all with an accuracy and speed that may surpass human capabilities. However, despite these advancements, AI systems will always lack human cognition's intuitive and subjective aspects. They will never experience the gut feelings that guide our decisions, nor will they comprehend the nuanced emotions that color our perceptions. The intuitive leaps of understanding, the sudden sparks of insight that characterize our thought processes and behaviors, will remain beyond their reach.

A significant strength of human intelligence is our ability to solve multidimensional problems creatively. Our intelligence, unlike machines, is remarkably flexible and multidimensional, sparking a creative spirit that any artificial entity cannot replicate. While machines and AI systems often excel in single-dimension tasks, where success or failure can be quantified within a specific parameter, the complexity of human cognition shines in multidimensional scenarios that necessitate a nuanced understanding of diverse factors. Unlike AI systems that typically operate within the confines of their training data, humans have the potential to synthesize diverse sources of information - weaving together strands of knowledge, cultural context, emotional cues, and personal values to devise innovative solutions. This enables us to view problems from myriad perspectives. This ability to synthesize diverse information allows us to see beyond the immediate problem, considering the wider implications of our actions and their potential ripple effects.

Our capacity for imagination and foresight sets us apart as well. By anticipating hypothetical scenarios and predicting the consequences of various actions, we can make informed decisions that consider each option's potential benefits and risks, a skill that is simply invaluable. Our brains are not just calculating machines; they are complex simulation machines, taking us beyond the boundaries of the present and enabling us to envision the future. While AI can process information and make predictions based on past data, it lacks the capacity for this kind of abstract and future-oriented thinking.

Another vital aspect of human problem-solving is our ability to communicate and collaborate effectively. This might seem like a simple capability, but it holds immense power. We are social creatures and have developed intricate social structures and languages, enabling us to share ideas, pool knowledge, and build consensus. This is one of our greatest strengths, and we can devise solutions far superior to what we might have come up with in isolation. This ability to orchestrate collective intelligence is particularly crucial in the face of complexity that demands diverse insights. Most problems we face are highly subjective, with no single easy answer, but a delicate act of balancing the pros and cons, making difficult choices by sacrificing certain benefits, comforts, and more. By drawing from diverse perspectives and expertise, humans can often arrive at superior, well-rounded solutions.

Aside from this, at the heart of human problem-solving is a spirit of creativity and adaptability that sets us apart from other species. Well-defined rules or patterns do not bind us; instead, we can think outside the box and devise new ways of addressing challenges. This creative mindset allows us to adapt to ever-changing circumstances and develop novel solutions that can transform our world for the better. When confronted with a problem, humans can approach it intuitively, incorporating empathy and creative thinking. While machines excel in repetitive tasks and those that adhere to established rules and patterns, humans thrive in ambiguity, where these rules are undefined or in flux. These qualities equip us to tackle nebulous, real-world problems that escape the strict confines of logic.

Just as a star shines brightest in the darkest night, the luminous facets of our humanity come to the fore as we navigate the dawn of the AI era.

Before delving into the strengths of machines and AI, let's look through some finer, often unrecognized, uniquely human skills in the modern workplace.

1. **Empathy and Emotional Understanding**: Humans have the innate ability to understand and empathize with the emotions of others. This skill allows us to connect deeper, provide support, and navigate social situations. While machines can recognize and analyze emotions to some extent, they lack the genuine emotional understanding that humans possess.

2. **Contextual Understanding and Nuance**: Humans are adept at considering context and recognizing subtle nuances in language, behavior, and situations. On the other hand, machines often struggle with understanding the subtleties and complexities of human communication and culture.

3. **Moral and Ethical Judgment**: Humans can make moral and ethical judgments based on various factors, including personal values, societal norms, and empathy. Machines, while able to follow programmed rules and guidelines, lack the inherent understanding of moral and ethical complexities that humans possess.

4. **Intuition and Gut Instinct**: Humans can rely on their intuition or gut instincts when making decisions or solving problems, often drawing upon a wealth of experience and knowledge that cannot be easily quantified. Machines, in contrast, rely on programmed logic and data-driven analysis, lacking the intuitive capabilities of humans.

5. **Artistic Expression and Aesthetics**: Humans have a unique capacity for creating and appreciating art in various forms, which often involves expressing emotions, conveying complex ideas, or evoking a sense of beauty. Machines can be programmed to generate art but lack the innate human understanding of emotions, aesthetics, and the cultural context that drives artistic expression.

Understanding these unique aspects of human capability is vital for us to remain relevant in the future workforce as AI systems' capabilities continuously improve and expand.

Under the blinking lights and seamless metal exteriors of machines and AI, some abilities far exceed our human capabilities in certain domains. Let's look at these machines' powers through the lens of Generative AI, specifically in the context of "mental tasks" and "cognitive capacity."

One of the most formidable advantages of AI and machines is their capacity to function as supreme librarians in vast information archives. They can instantaneously access and retrieve facts from colossal databases, managing information at a scale that is simply beyond human reach. This unparalleled access to massive memory resources gives machines a unique edge in tasks requiring speed and accuracy in data processing.

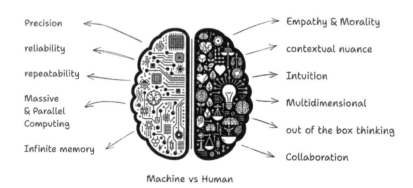

Machine vs Human

In today's era of big data, information accumulates in such large volumes that it is humanly impossible to sift through it all. This is where the second strength of machines comes into play - their pattern recognition abilities. Machine learning algorithms can scan through mountains of data in the blink of an eye, identifying subtle patterns and correlations that might slip past even the most observant human analyst. Tireless and meticulous digital detectives excel in fields where swift and accurate data analysis is vital, such as finance, healthcare, logistics, and many more, spotting patterns and trends.

Machines, in essence, are like master puzzle solvers, piecing together complex data sets to reveal hidden insights and knowledge.

Machines are the unchallenged champions for tasks that necessitate rapid data processing. Their abilities to work tirelessly, process information at lightning speed, and deliver consistently accurate results render them invaluable assets in today's data-driven world. Their vast memory and flexible access to computing power provide an undeniable edge. In a world where information is the new oil, machines are our ultimate refineries, transforming raw data into invaluable insights and decisions.

Adding to their advantages is the steadfast consistency and reliability of machines. Unlike humans, who are susceptible to errors, fatigue, or bias, machines can perform tasks with unparalleled precision and accuracy time and time again. They operate with Olympian-level precision and reliability, whether it's the day's first task or the thousandth, a mundane job, or a complex calculation.

Dissecting the man-machine capabilities from this lens gives us a simple strategy to thrive in the era of machines ascending in cognitive capabilities. Increasing our footprint in domains of "Uniquely Human Qualities" and learning skills to use the ever-improving AI tools effectively is an effective way. Transforming ourselves into "Rule makers" and using machines to work with these rules is a way forward for humans to lead the transformation. By leveraging the strengths of both entities, we can achieve a level of performance that far exceeds what either could accomplish alone.

In this dynamically shifting ecosystem, where the lines between biology and technology blur, digital neurons intertwine with human intellect. The question isn't about the superiority of one over the other but about synergistically harnessing the unique capabilities of humans and machines. Certain questions beckon our attention as we steer our course through this unfolding narrative. How do we convert this understanding into actionable strategies in the real world? In what realms will AI assist us in pioneering groundbreaking strides? Where will the intrinsic human skills remain paramount, untouched by the cold precision of AI? And perhaps, most pertinently, where will AI threaten to cast a looming shadow over the human workforce?

In this context, we will explore the current job landscape through a distinctive prism — the degree of 'machine skills versus human skills' used in every job. This approach will help us design a "Man-Machine Map" matrix that illuminates the various shades of our technological era: the complementing, the competing, and the non-invasive characteristics. It will serve as our compass, guiding us through the intricacies of this dynamic ecosystem, delineating where human ingenuity will remain sacrosanct, where AI will forge new frontiers, and where a harmonious convergence of the two will drive exponential growth.

Man and Machine Map Framework

Let's create a simple two-dimensional graph, with one axis representing the number of human skills and the other representing machine skills involved in a job in today's professional landscape. We call this the "Man-Machine-Map". Here's how it might look:

1. **X-Axis: Degree of Machine Skills Involved** This axis ranges from low to high, representing the extent to which machine skills are involved in a profession.

2. **Y-Axis: Degree of Human Skills Involved** This axis also ranges from low to high, illustrating the extent to which human skills are involved in a profession.

The intersection of these axes gives us four quadrants, each representing a unique combination of machine and human skill utilization. Let's explore these quadrants, where strategies to navigate this impending job loss due to AI will become obvious.

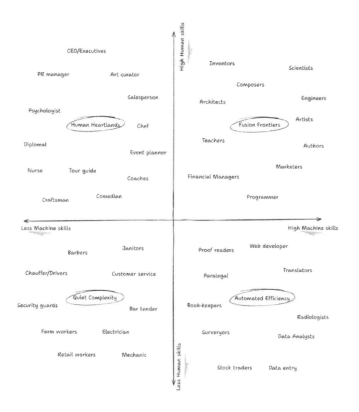

MAN - MACHINE - MAP (3M FRAMEWORK)

The journey to understanding human potential in contrast to artificial intelligence begins unexpectedly, not with the roar of machines but with the whisper of the human heart. In the landscape of skills and professions, we'll first venture into the "Human Heartlands," also known as the quadrant of "Empathetic Intuition." This is Quadrant II, a realm populated by professions requiring a deep wellspring of human skills, with a lesser need for machine capabilities.

This is the landscape of the therapists, teachers, social workers, writers, and artists. It's the realm of the musicians, chefs, diplomats, and coaches. Their tools are not just computers and algorithms but words, emotions, creativity, and human connection. It is an arena where human skills take center stage, the crux of our humanity - our ability to understand and empathize, our capacity for nuanced judgment, our potential for leadership, our inherent creativity, and our social and cultural understanding. In all these professions, there's complexity and nuance in human behavior and emotion that cannot be reduced to simple rules or algorithms.

Take, for instance, the profession of a school teacher. Some aspects of their job require machine skills, such as using digital tools for lesson planning or grading. However, the core of their work requires human skills that go far beyond that. They must understand and respond to each student's needs, inspire a love of learning, manage a classroom, and communicate effectively with parents and colleagues. They need creativity to make lessons engaging and adapt their teaching methods to suit different learning styles. No algorithm could replicate the nuanced understanding and empathetic response or the connection with the students a teacher brings to their classroom.

Or consider a psychotherapist whose work is deeply rooted in empathy and understanding the human psyche. They navigate the labyrinth of human emotions, pick up on subtle cues in body language and speech, and build a relationship of trust with their patients. They guide people through trauma, grief, mental health struggles, and life transitions, a role that demands a depth of empathy, intuition, and emotional resilience that can't be coded into a machine.

In art, professions such as actors, musicians, or painters require a depth of emotional expressivity and social understanding that is impossible to code into an algorithm. An actor doesn't just deliver lines; they breathe life into a character, understanding their motivations, backstory, and the subtle nuances of human behavior that bring authenticity to their performance. A public relations specialist, another professional in Quadrant II, is tasked with protecting and improving their clients' public image. This requires understanding not just the client's needs but also the perceptions and sentiments of the public, the current socio-political climate, and how to communicate effectively within this context. AI can analyze trends and generate communications, but the human touch needed to navigate these complexities and make strategic decisions is far beyond its capabilities.

The realm of this quadrant stretches across various levels of sophistication in our earlier Skill-Scope Canvas. At the upper echelon of sophistication, we find roles like CEOs, enterprise executives, judges, and diplomats. For instance, a CEO is tasked with steering a company's strategic direction, leading its people, understanding their motivations, and nurturing a culture fostering growth and innovation. They must navigate the complexities of human behavior within an organization while managing the external pressures of a fast-paced, ever-changing business environment. Similarly, judges and diplomats must exercise nuanced judgment and employ a deep understanding of social, cultural, and political contexts. Their decisions are shaped not only by the letter of the law or the explicit terms of treaties but also by the spirit of justice and the nuanced dynamics of international relations. It's a role that demands a high degree of human skills, a depth of understanding, and a breadth of perspective beyond AI's grasp.

At the other end of the sophistication spectrum, we find roles like salespersons, nurses, tour guides, and social workers. While these roles might not have the same level of strategic or decision-making responsibilities as the aforementioned ones, they are highly human-centric and demand a robust suite of human skills. A salesperson's success hinges on their ability to connect with customers on a human level, understanding their needs, desires, and reservations. Nurses provide medical care and emotional support to patients and their families, navigating the vulnerability of human health with compassion and empathy. Tour guides don't just share facts and figures about historical sites or natural wonders; they narrate stories, spark curiosity, and create a human connection with their audience. Social workers step into the labyrinth of social issues, armed with empathy and understanding, to help individuals navigate complex personal and societal challenges.

While these roles might not require the same level of sophistication as a CEO or diplomat, they are nonetheless grounded in human skills that machines can't replicate. The empathy of a nurse, the storytelling ability of a tour guide, the salesperson's knack for persuasion, or the social worker's capacity for compassion cannot be automated.

In essence, Quadrant II celebrates our human nature across all levels of sophistication. It underscores the idea that some tasks and professions should always remain human. Whether we're talking about CEOs or salespeople, the value they bring is inherently human and irreplaceable. Their roles demand a sophisticated understanding of human nature and society, the ability to navigate ambiguity and incomplete information, and a depth of empathy and creativity that no machine can replicate. These deeply human qualities make the professions in Quadrant II not only resistant to automation.

Currently, and likely for the foreseeable future, these roles represent a haven that AI and machines are far from reaching. It should remain a quadrant where humans should not let AI perform, driven by short-sighted economic benefits, even if they become feasible. As we continue to navigate the evolving landscape of skills and professions, this quadrant serves as a reminder of the enduring power of the human touch in a digitized world.

As we continue our exploration of the landscape of human and machine skills, we find ourselves entering a paradoxical domain: the third quadrant. This is a land less traversed, a domain where the tasks don't demand the higher echelons of uniquely human skills, nor are they tasks that machines and AI have yet conquered. It's an intriguing place, brimming with an array of jobs that, on the surface, might seem simplistic but, beneath that facade, hold a complexity that keeps them elusive to the grasp of automation. Let's call this domain the "Quiet Complexity".

Within this Quiet Complexity quadrant, we encounter professions that defy easy categorization. At first glance, they may seem to require low human and machine skills. Still, upon closer examination, these tasks reveal a multidimensional and physically nuanced complexity. We find the professions that involve a direct and physical engagement with the world - the gardeners, the plumbers, the electricians, the carpenters, and the chefs. These are roles where humans continue to outshine machines, not because of our cognitive superiority but due to our physical talent, adaptability, and ability to navigate the unpredictable, complex nature of the physical world.

While these professions don't necessarily require the high level of empathetic intuition found in the Human Heartlands, they still necessitate a certain degree of human touch. These tasks involve complex physical skills and an understanding of real-world materials that are hard to automate.

A gardener, for example, must know when to prune, how to combat pests, and when to plant specific species. They must understand the sun, soil, water, and weather interplay. They need to dig, plant, weed, and harvest physically. The understanding required here is more tacit and experiential than explicit and codified. Likewise, a plumber or an electrician deal with physically complex and highly variable systems. No two houses are identical in their plumbing or electrical layout. They often need to physically access tight or awkward spaces and use their hands in ways that are difficult to automate.

These tasks may seem ripe for automation, but the economic reality and physical complexity make them unattractive targets for AI and machines. As discussed earlier, automating these tasks would require robots with physical skill and adaptability. This will not be a reality for a long time, inhibited not just by technological sophistication but also by the market forces of economic viability.

As we delve deeper into Quiet Complexity, we uncover the subtle and often overlooked dimensions of work that are neither dominated by machines nor saturated with higher-order human skills. These jobs illustrate a different but equally significant facet of the human-machine contrast, showcasing tasks that remain firmly in human hands due to their physical complexity, economic unattractiveness for automation, and the intrinsic human satisfaction they provide.

The landscape of the third quadrant, the "Quiet Complexity," is indeed shifting with the rapid advancement of technologies with AI. Many tasks that would have comfortably nestled here a few years ago are now being pushed into the fourth quadrant, where machines have become increasingly adept. This has been especially true for jobs requiring a high cognitive load but not necessarily involving a significant degree of human-to-human interaction or the higher echelons of uniquely human skills. You can imagine any profession that doesn't involve much human interaction and mental nature, which was placed here before the AI era. Consider jobs of "Data entry," "Bookkeeper," or "Data analysts" that have been silently pushed to the 4th Quadrant.

As cognitive tasks become increasingly automated, the "Quiet Complexity" has been honed into what we might aptly call the "Tactile Territories." In this realm, the roles are decidedly practical and hands-on. The labor here is often physical, requiring the kind of nuanced manipulation of the material world that machines have yet to master.

The Tactile Territories remind us that while we often celebrate humans' cognitive capabilities—our ability to think, reason, and create—there is also something fundamentally human about our ability to interact with and manipulate the physical world. Even as AI technology advances, these tactile skills keep a significant portion of the world of work firmly in human hands.

As we journey from the realm of "Empathetic Intuition," and "Quiet Complexity" continues, our path leads us to a unique crossroads, a fusion of the uniquely human and the power of the machine. This is Quadrant I, a vibrant marketplace of synergy and integration where high human skills and high machine skills converge. This is the land of opportunity where technological prowess complements human intuition, creativity, and empathy to unlock new frontiers.

In this realm, machine skills aren't replacing human ones; they're a powerful extension, enhancing our capabilities and amplifying our reach. Here, we find software developers, aerospace engineers, cybersecurity experts, financial analysts, architects, educators, and even artists - professionals who leverage advanced technologies to augment their human abilities and unlock unprecedented possibilities.

One example to illustrate jobs in this landscape is the aerospace engineer, wielding the power of generative AI to reach into the cosmos. With each spacecraft design, they're not merely drafting a blueprint but envisioning humanity's place among the stars. Generative AI becomes a dynamic ally in this process, enabling engineers to navigate through countless design permutations at an unthinkable pace. Yet, their deep understanding of physics, creative problem-solving, and ambitious human spirit guide this high-tech tool to realize visions of efficient, next-generation spacecraft.

Architects in this quadrant aren't just crafting buildings; they're sculpting the future landscapes of our cities. Generative design offers them a playground of possibilities, enabling them to explore daring, out-of-the-box designs. The AI isn't the architect here; it's the enabler, a sophisticated drafting board that can instantly materialize a thousand design paths. The architect's vision, aesthetic intuition, and understanding of human needs and experiences breathe life and purpose into these structures.

The software developers, the code poets of the digital era, utilize AI to elevate their craft. While AI can automate routine tasks, handle debugging, and even optimize code, the developers' logical reasoning, creativity, and understanding of human needs birth innovative software solutions. They use AI not to replace their skills but to create new programming paradigms that promote efficiency and simplicity, offering higher-level abstractions and allowing them to focus on solving more complex, creative problems.

In finance, managers are partnering with AI to gain a firmer grip on the vast financial landscapes. AI algorithms sift through enormous datasets, spotting trends and making forecasts at superhuman speed. But it's the financial manager's deep understanding of market psychology, their strategic insight, and their capacity for ethical judgment that make sense of this information, helping them craft strategies and plan resources effectively.

Even music composers are teaming up with AI to explore new melodic territories. AI can generate novel musical sequences, offering a sandbox of auditory possibilities. Yet, the composers' artistic intuition, emotional expressivity, and profound understanding of the human response to music shape these raw AI-generated notes into meaningful, moving symphonies.

Quadrant I is an exhilarating realm where human insight and machine intelligence form a powerful alliance. Here, we see how our uniquely human skills – creativity, critical thinking, and understanding of rules – remain indispensable, even as we leverage sophisticated AI tools. It's not about choosing human skills over machine skills or vice versa, but understanding how they can work together and enhance each other.

In this quadrant, humans' strengths are their creativity and the ability to think outside the box, imagine new possibilities, and turn them into reality.

In this realm, AI doesn't lead; it follows. It follows the human vision, intuition, and touch. It's a tool, a companion, a canvas on which we paint our dreams. It doesn't limit our potential but amplifies it, helping us achieve what was once impossible.

As AI technology evolves, Quadrant I will become an even more crucial part of our professional landscape. This is the land of

"Integrated Innovation," where humanity and technology join forces to shape a future beyond our wildest dreams. These are also jobs where AI could play a complementary role, and the jobs could be lost due to the efficiency of a skilled person who can do a job of 10 people with the support of AI tools. This is a land of caution as the line between the ability to use these tools and becoming irrelevant due to the effective multiplication brought in by these tools will become very thin.

Our exploration of the human-machine skill continuum ends with the fourth quadrant. It's a fascinating terrain where machine skills are highly utilized, and the requirement of uniquely human skills is relatively low. This quadrant has perhaps seen the most dynamic change, its boundaries constantly shifting and expanding as machines' cognitive capacities grow exponentially.

Welcome to the domain of "Automated Efficiency," a living testament to the relentless march of technology and its profound impact on the world of work. It's a realm ruled by machines, where repetitive tasks, pattern recognition, and rule-based procedures reign supreme. This is the landscape where technology has made its most dramatic encroachments, transforming fields once seen as completely human. This is where we see the story of tasks once performed by human hands now being executed by machines with efficiency and precision that far outstrips human capabilities. As technology advances astoundingly, this quadrant continuously expands, constantly reshaping the boundaries of what is possible with AI and machine learning.

There's an almost Darwinian quality to this shift. As we trace the path of professions such as weavers, metalworkers, or cart pullers, we see jobs once deemed indispensable being automated into oblivion. These tasks, once requiring the sweat and toil of hundreds, even thousands of workers, are now completed by machines under the watchful eyes of a few supervisors. Weavers, once a mainstay of the textile industry, have largely been replaced by machines that can produce fabric at a rate no human could match. Once revered for their skill in shaping metal into useful tools and ornate artworks, metalworkers can't compete with automated machines that can precisely replicate their work. Cart pullers and knockers, whose roles have all but vanished in the modern world, were early casualties in the steady encroachment of machines into the realm of human work.

Today, we are witnessing a similar trend, with jobs that once seemed safe in the domain of humans being nudged into the fourth quadrant. They have found a new home in this quadrant, and the cognitive abilities of AI are expanding to encompass these areas, too. Data analysts, copywriters, social media marketers, web developers, and more find their roles increasingly automated. Sophisticated AI can now easily sift through vast datasets, craft persuasive marketing copy, design eye-catching websites, and even manage social media accounts with an efficiency far surpassing human capability. Those currently employed in these fields stand on precarious ground, their jobs threatened by an unrelenting tide of technological advancement.

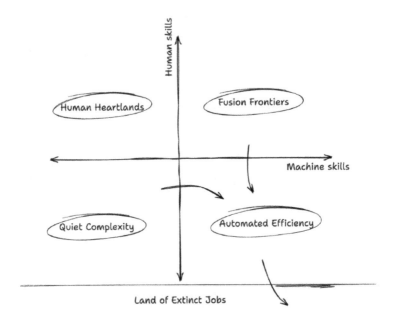

Proofreaders, once the final bastion of human skill in the publishing process, are increasingly finding their roles usurped by software that can spot grammatical errors, typos, and even issues with syntax.

Similarly, radiologists, who were once solely responsible for reading and interpreting medical imaging, now work alongside sophisticated AI that can detect anomalies with incredible accuracy.

Translators, too, once relied solely on their intricate understanding of languages, are now aided by powerful software that can instantly translate text from one language to another. Paralegals and bookkeepers find their roles increasingly automated, with software now capable of sorting through legal documents or managing accounts with minimal human intervention.

Stock traders, once the masters of Wall Street, compete with automated trading systems that can analyze market trends and execute trades at lightning speed. Medical transcriptionists and data entry clerks, too, are finding their roles changing, with voice recognition and data processing software becoming more accurate and reliable.

Surveyors, whose profession once required extensive fieldwork and hands-on measurements, are now assisted by drones and sophisticated mapping software that can conduct surveys with pinpoint accuracy from the comfort of an office. These are just some illustrations of some jobs that fall into this category. They give an idea of some types of jobs and "why" they fall into this category, but they are not an exhaustive list.

An unsettling question arises here - what happens when the tasks we perform and the skills we've honed are automated? What happens when the job you've spent a lifetime perfecting is suddenly done faster and more accurately by a machine? And more importantly, where does that leave us as humans? What value do we hold in a world increasingly dominated by machine efficiency?

It's a chilling thought and one that many might prefer to avoid. But avoid it, we cannot. The relentless gears of progress grind on, and standing still is not an option. As technology evolves, so too must we adapt. This relentless shift poses a significant challenge for those currently employed in these domains. For them, the future may seem uncertain as they face the very real risk of their roles being automated. This upheaval underscores the urgent need for workers in these fields to adapt and prepare for a future where their current roles may be radically transformed or even rendered obsolete.

Having understood the nature of human jobs from this perspective, it's obvious to foresee how our roles will be affected. The most effective strategy is to position yourself as "Master of machines" and "not be Mastered by machines"

Strategies To Navigate Job-Loss

Yet, within this stark reality, these frameworks – the SSC and the 3M – offer clarity and guidance. They dissect the mechanics of job disruption, stripping away the hype, uncertainty, and hysteria that so often cloud our understanding of AI's impact. They allow us to see the transformation not as an undefinable, indomitable force but as a series of navigable shifts. The solution, while conceptually straightforward, can seem intimidating in its magnitude: we must journey beyond the borders of the 'Automated Efficiency' quadrant, seeking refuge and reinvention within the remaining three. In this voyage, we're not simply drifting aimlessly; our compass is set towards a higher level of sophistication.

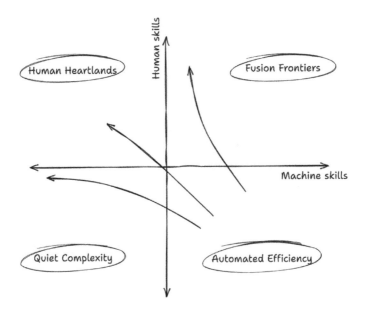

Given the structure of our four Quadrant models, the first obvious approach is to seek higher ground, or more specifically, to move higher up on the human skill axis within Quadrant 1. At the crux of this migration is the development of skills and capabilities that are uniquely human, traits that machines, regardless of their sophistication, cannot replicate. These skills, which lie at the intersection of creativity, critical thinking, emotional intelligence, and complex problem-solving, form the bedrock of our uniquely human potential. These skills are not constrained to be relevant in specific domains; they are generic skills that vary in applicability across domains. However, these are some unique aspects that underline the direction of professional competency. By honing these skills and building proficiency, individuals can ascend the human skills axis, effectively distancing themselves from the reach of automation. The higher we climb, the more indispensable we become, as these skills are not only less susceptible to automation but also in growing demand in today's economy.

However, this isn't a one-size-fits-all solution. The path will look different for everyone, shaped by personal interests, innate abilities, and the socio-economic context. Some might deepen their expertise in a specific domain, while others prefer to broaden their skills across different fields. Some might transition to roles that rely heavily on human skills, such as social work, education, or leadership positions. While these skills might seem generic, abstract, and intangible, they can be nurtured and developed to work together with advancing AI technologies. This journey of professional evolution might appear to be a Herculean task. Still, it can be guided by two age-old yet infinitely powerful paradigms - '*Original Thinking*' and '*Value Creation*.' These paradigms have always been central to human endeavor. Still, in this era of AI disruption, they will become our guiding stars.

Rule Makers vs Playing by the Rules

Understanding the fundamental difference between the *learning process* of humans and machines is the key to charting our strategy. This distinction, while subtle, deeply impacts how we and our machine counterparts interact with, understand, and navigate the world around us.

Let's revisit the concept of 'factors' and 'components' as discussed in an earlier chapter. We marveled at how machines, when exposed to vast amounts of data, learned to detect the underlying probabilistic associations or 'latent factors' and piece together the puzzle of 'components' to form an output. Whether it's an image, a piece of code, an architectural design, or a passage of text, AI systems, much like diligent archaeologists, excavate the game's rules from the mountains of data they sift through.

Imagine, if you will, a giant jigsaw puzzle. Each piece represents a 'component,' fitting together to create a comprehensive 'output' image. AI systems learn by examining thousands, millions, and even billions of completed puzzles. By absorbing this vast sea of data, they perceive the patterns and connections underlying these completed images - the 'latent factors' or the game's rules. Consider an 'AI architect.' Through extensive exposure to blueprints and designs, it starts to discern the 'latent factors.' It shows that the foundation always comes first, that a window is part of a wall and generally faces outward, and so on. This understanding is largely statistical, rooted in the observation of countless examples.

In other words, AI deduces the game's rules by meticulously observing the play.

Humans, on the other hand, operate in the reverse order. We first learn the 'rules of the game.' We construct sentences by understanding the basic principles of grammar; we don't start constructing buildings before knowing what a foundation is and why it's important. They delve into the semantics of each 'factor,' decoding the logic that drives the placement of the 'components.' We grasp the philosophy, the 'why' behind the 'what.' With this framework of understanding in place, we create outputs by assembling various 'components' guided by these rules.

This learning paradigm is mirrored in the way humans master language. We don't just mimic phrases and sentences we've heard thousands of times. Instead, we learn the grammar - the rules of language - which allows us to create coherent sentences from individual words. Even with a limited vocabulary, we can create endless sentences, thanks to our understanding of grammatical rules.

Imagine a child learning a language. They don't start by reading entire books and then deriving grammar. Instead, they begin with the fundamentals—phonemes, words, simple sentence structures—and gradually layer on more complex grammatical rules. As their linguistic prowess grows, they can craft intricate narratives, play with metaphors, and captivate listeners eloquently.

This dichotomy doesn't just underscore the differences in how humans and machines process information. It also provides insight into how we can best use AI systems. With its power to process vast amounts of data, AI can unearth patterns and associations far beyond human capacity. The human ability to understand rules and their implications, to draw on subjective experiences, and to make intuitive leaps remains unmatched. We will draw further upon this in the next chapters.

The narrative of AI systems and human cognition opens a new chapter as we grapple with the nature and boundaries of 'the box' – the realm of knowledge and possibilities within which humans and machines operate. The crux of the matter lies in the understanding that machines, for all their prowess, function 'inside the box' of data they are exposed to. They navigate this labyrinth with impressive efficiency, detecting patterns, extracting rules or latent factors, and producing outputs that either comply or, at best, marginally deviate from the observed 'rules of the box.'

The box that AI systems operate within is vast, furnished with countless bits of data, and powered by near-infinite memory and computational ability. Indeed, there is immense potential to unlock just by mastering the art of juggling components within this mammoth box. Yet, the AI's creations, no matter how sophisticated, are confined within the parameters of the extracted rules. They are remarkable renditions drawn from an ocean of observed examples but still inherently bound by the confines of their learning.

In contrast, humans' approach to cognition is altogether different. Humans, in stark contrast, are rule-definers and rule-breakers. They have the unique ability to learn the rules and modify, redefine, and invent new ones. This is where the human intellect truly comes into play. We can expand the box's dimensions, push the boundaries, and pave the way for new combinations and possibilities stemming from newly defined or altered rules.

An artist, for example, may learn the 'rules' of color and form but then choose to disregard them, creating abstract art that bears little resemblance to observed reality. A scientist may learn the existing theories and laws but then propose new hypotheses that push the boundaries of our understanding. This ability to redefine the 'rules of the box,' to create and innovate outside established patterns, is a distinctly human trait.

Understanding these differences in cognition is paramount, for we might be swept up by the velocity of technological progress and struggle to keep track of every new development. However, these foundational principles—the dichotomy between human and machine learning, the confines of 'the box,' and our respective abilities to manipulate the rules—will be our compass in this uncharted territory.

Original Thinking

Expanding the box or stepping into the 'Original Thinking' arena is akin to becoming an explorer in an endless landscape of imagination, a vast and vibrant territory uncharted by artificial intelligence. In the realm of AI, impressive as its capabilities may be, it remains confined within the confines of its training data, within the perimeter of its 'box.' The AI juggles the contents of this box with astonishing skill, performing feats of recombination that can seem nothing short of superhuman. But no matter how big or multifaceted the box may be, AI can never venture beyond its limits. It's here that 'Original Thinking' takes center stage.

Original thinking is about extending the boundaries of this box, expanding the dimensions of possibility. It is a uniquely human trait – the ability to take a leap of imagination, venture into the unknown, and return with something entirely new that did not exist within the data the AI was trained on. It's about introducing novelty, creating something from nothing, a spark ignited from the ether of the unimagined.

Consider the process of invention. The creation of the telephone by Alexander Graham Bell, the formulation of the theory of relativity by Albert Einstein, and the discovery of penicillin by Alexander Fleming were acts of original thinking. No amount of recombination of existing data would have led a machine to these insights, for they required a leap into the unknown, a step beyond the familiar, and a dive into the depths of intuition. Not to set a very high bar, but original thinking doesn't just imply that one needs to be an exceptional scientist or inventor. It's also about coming up with unique solutions to everyday problems.

And it's not just about expanding the box but also about changing its shape entirely. It's about defying conventions and breaking norms, about seeing the world not as it is but as it could be. It's about challenging the status quo and daring to think differently. In this case, the' box' isn't just a data repository but also a matrix of established paradigms, traditional perspectives, and widely accepted norms.

AI sees the world as it is; Original Thinking envisions it as it could be.

As we navigate the age of AI, when the dust of frenzied development eventually settles, those with the ability to think originally, create new content, and venture beyond the known and familiar will be in high demand. Original thinkers are the pioneers of progress, the vanguards of innovation, and the architects of the future. They are the ones who will introduce new elements into the 'box,' providing AI with fresh data to juggle, thereby driving progress. Original thinking is not an exclusive ability of a gifted few; it's a trait that can be nurtured and developed. The emergence of AI has not diminished the value of original thinking; on the contrary, it has magnified its importance.

Imagine you're an artist, standing before a blank canvas with a paintbrush. The world around you ceases to exist as you surrender to the rhythm of your creative heartbeat. The art you create isn't simply a mere depiction of reality; it's a deeply personal expression of your thoughts, emotions, and experiences - it's a window into your soul.

In contrast, artificial intelligence, equipped with advanced algorithms and access to countless works of art, can generate various artistic outputs. It can mimic the brush strokes of Van Gogh, the color palettes of Matisse, or the surrealism of Dali. It can blend these styles into entirely new forms, creating outputs that can seem innovative, fresh, and even breathtaking. Yet, these AI-generated works of art are essentially a recombination of existing styles. They're an amalgamation of elements previously defined by human artists. In essence, the AI's 'creativity' reflects the creativity of the artists whose works it has been trained on.

As AI-generated art becomes more widespread, it will have myriad applications. We may see AI-created graphics embellishing our advertisements, enriching our virtual worlds, and adorning our everyday objects. Personalized AI-created art may become popular gifts, and AI-generated illustrations may enhance creative stories. But as AI-generated art becomes increasingly commonplace, something remarkable will happen: the value of truly original human art will rise. In a world awash with AI art, the art that sparks interest will be the one that provides something different, something authentic, something human.

For you, as an artist or graphic designer, this presents an unprecedented opportunity. Your ability to create new styles, express unique perspectives, delve deep into the human psyche, and portray emotions in ways that machines can't mimic will be your trump cards. Art can encapsulate the nuances of joy, the depth of sorrow, the intricacies of love, or the sting of betrayal. You can produce art depicting new styles, provoking new thoughts, inviting introspection, and challenging norms.

In the age of AI-generated art, you are not the canvas that AI paints on; you are the artist who expands the box of creativity, reshaping it with each brush stroke.

And so, as an artist in the age of AI, your task is not to compete with the machine but to complement it. Use AI as a tool to augment your creativity, not as a competitor to intimidate it. Remember, AI might create a thousand sunsets, but it will never know the sun's warmth on its silicon skin. It might depict a thousand human faces, but it will never understand the story behind their smiles or the sorrow in their eyes. And that is the artist's domain – to delve into the depths of human experience and express it in ways that resonate with other humans.

This same principle can be applied to other domains. One can think of how this is relevant in marketing, writing, filmmaking, law, or even software development.

Original thinking demands a reconfiguration of our learning strategies to accommodate the importance of breadth, or "multidimensional understanding," and depth, or "deep conceptual understanding from a first principles level." This dual approach requires comprehensive knowledge across multiple domains and an in-depth understanding of the chosen area of expertise. It's like becoming a Renaissance person with the precision of a master craftsman.

In a world saturated with information, we're often seduced by the ease of superficial learning. YouTube tutorials, quick-read summaries, and online crash courses offer us an illusion of mastery. The temptation is to equate familiarity with expertise. But true mastery goes beyond this superficial knowledge. Deep understanding, or expertise, is about truly comprehending the heart of a subject, peeling back the layers to explore the why, not just the what or the how. This often involves breaking down complex concepts to their most fundamental truths and building up from there.

Let's continue with the example of marketing. A deep understanding would involve knowing more than just how to operate analytics tools or execute social media strategies. It would require understanding the basic principles of human behavior, how and why people make buying decisions, and the psychological triggers that can influence them. A marketer with this depth of understanding can craft campaigns that truly resonate with consumers and feel personal, engaging, and real. They're not just following a set of best practices but creating strategies based on a deep understanding of human nature.

On the other hand, a breadth or multidimensional understanding involves knowing a range of different fields or areas. This kind of 'T-shaped' knowledge, combining depth in one field with a breadth of understanding across many, allows for the cross-pollination of ideas, sparking innovation and creativity. A marketer with a broad understanding might draw on insights from psychology, design, data science, and cultural studies to inform their strategies. It might involve studying behavioral economics to understand consumer decision-making, diving into design principles to understand the impact of visual elements in advertising, or even learning about cultural nuances that influence buying behavior. This breadth of understanding allows us to see connections, adapt and innovate when faced with new challenges, and apply our knowledge innovatively.

Both depth and breadth are critical to original thinking. While depth provides the foundation for a solid understanding of principles, Breadth fuels creativity and the ability to draw on diverse knowledge and perspectives to generate new ideas. Together, they make us better problem-solvers, more creative thinkers, and, ultimately, more irreplaceable in the age of AI.

'Original Thinking' is not just about inventiveness but the ability to connect dots to synthesize diverse information streams into something entirely new. It's about discerning patterns where others see chaos, asking questions that haven't been asked, and daring to venture where others haven't. In original thinking, we find the roots of innovation and the seeds of transformation.

Now, let's shift our focus to the realm of software engineering. The proliferation of high-level programming languages, libraries, and frameworks has undoubtedly made coding more accessible. You can now build a functional website with a few lines of Python or develop a mobile app using a visual interface—no deep understanding of computer science is required. However, these advancements also make certain coding tasks more susceptible to automation. Consider a web developer who might be proficient at tying together APIs, making impressive front-end visuals, and even producing functional software. They may not survive in the long run. As AI tools evolve, many of the routine tasks associated with these areas will likely be automated.

This brings us to the paramount importance of depth in software engineering. One must go beyond knowing how to code to excel and be irreplaceable in this area. Understanding computing principles— from the silicon level up—gives engineers a robust foundation to build and innovate. It's crucial to understand the fundamental principles of computer science, how bits and bytes travel through the hardware, how data structures and algorithms work, and the nuts and bolts of computer architecture and operating systems. This fundamental knowledge imbues them with a versatile skill set that enables them to think beyond the code and address issues like efficiency, scalability, security, and maintainability. It allows them to anticipate future requirements and design resilient and adaptable systems.

Moreover, having such deep knowledge can allow software engineers to develop new programming languages or systems that can transform the tech industry. They can design new architectures that are more efficient or that meet specific needs in the market. Just think about the inventors of revolutionary technologies like Linux, Python, or even the World Wide Web. These were all created by people with a deep understanding of software and web development principles, allowing them to make contributions that transformed the industry and are still relevant today.

These examples illustrate the tenets of "Original Thinking," and this can be applied to any relevant field. Now comes another important question.

> **We can "Think originally" in all possible directions. It can be a wide-open landscape that proves pointless. How do we channel this force to be relevant in the economy? As we have seen earlier, a crucial factor in the evolution of technology is "value creation." This latent aspect driving the "market forces phenomenon" applies to technology and us as individuals. Once we understand the real value-creation aspects in our respective domains, we can learn to use the AI tools to effectively deliver value.**

Art of Value Creation

Value and value creation are relevant and central to understanding our future in an AI-driven world. Certain roles may become less relevant as artificial intelligence (AI) evolves, but this doesn't spell doom and gloom. On the contrary, this shift will create new avenues for value creation, requiring our distinct human attributes.

Simply put, "value" signifies the impact, benefit, or importance of something. In business, it's the advantage or benefits a product or service offers compared to its cost, essentially fulfilling needs or enhancing experiences. Value creation becomes a quest to magnify these rewards, whether innovating a product, refining a service, or fostering novel ideas.

Imagine value as a lighthouse. While the sea of life is unpredictable and tumultuous, with waves of uncertainties, a lighthouse provides direction, clarity, and safety. It's a guiding light, ensuring ships find their way home. Similarly, value is about offering clear benefits and addressing specific needs, acting as a reliable guide in the marketplace.

Now, let's shift gears to value creation. It's like the role of a talented magician at a party. What does a magician do? They pull a rabbit out of a hat, right? But it's not just about the rabbit or the hat; it's about the awe, wonder, and joy the magician creates for the audience. That's value creation—making something from nothing or transforming the ordinary into the extraordinary, like turning a hat into a home for a fluffy bunny. It's not about the objects; it's about the effect on the crowd.

Let's visualize "Value and Value creation" with the metaphor of a potter at his wheel. The potter starts with a simple lump of clay, which holds potential but little value. As he shapes and molds it, the clay forms a pot, a vase, or a dish. With every wheel spin, the potter's careful hands sculpt more value into the object. Finally, after going through the fire of a kiln, it emerges as a beautiful, useful piece of pottery – something of significant value. In the business world, it's a similar story. A startup may begin with nothing more than an idea, just like a lump of clay. As the concept develops, the potential increases. The idea is shaped and molded into a product or service while adding layers of value. When it finally enters the market and addresses a need or solves a problem, the startup creates the ultimate value – it becomes valuable to its customers.

The concept of value is also akin to a chameleon, changing colors based on the context, timing, and perceived benefits. This dynamic nature of value makes it a cornerstone in the economic, social, and personal world. Imagine value as an elusive genie hidden away in a magical lamp. Not everyone sees the lamp, but those who understand its utility. This lamp can grant them their most desired wishes - just like a product or service can fulfill a need or want. But this lamp doesn't hold the same value for everyone. Some might be intrigued by its aesthetic beauty, its mysterious aura might spellbind some, and others might be desperate for the wishes it can grant. It's all about the right person finding it at the right time.

Like the genie in the lamp, the value of a product, a service, or even an individual's skill set morphs depending on the situation, just like a pair of woolen socks that seem useless on a warm beach but become a lifeline in freezing temperatures. The importance of understanding this ever-changing nature of value cannot be overstated.

The same applies to the job market. As an employee, your skill set is like a key. On its own, the key might seem insignificant. But its value skyrockets if it can open a door that no other key can. Especially with the advent of AI tools that are getting more sophisticated every day, one needs to identify the unique value they bring to help them stay relevant constantly.

In today's world, where new AI systems are released almost every day, it's easy to lose sight of this line, separating a skilled operator from the tools used. Especially in this era, knowing this difference will be "the" factor distinguishing between being indispensable or irrelevant. Let us understand this difference with a real-world illustration.

A case in point is the role of a software developer. The same quantum of work will inevitably be delivered with significantly smaller teams.

A developer, traditionally seen as a 'coder,' was often valued for his or her technical prowess and proficiency in certain programming languages, let's say C++, Python, or Java. However, the reality is more nuanced. The core value lies not in their mastery of a programming language but in their ability to leverage that skill to create solutions that address business needs. Consider the process of developing a mobile application. A developer might code an excellent, bug-free application that works seamlessly. However, its true value is not determined by the code quality alone but by its utility. Does it cater to the needs of the users? Does it solve a problem?

When a developer sits at their workstation, they're not just constructing lines of code; they're essentially building digital bridges, creating connections, and removing barriers. They're contributing to a product or service that could solve a pressing problem or cater to a market need. A developer's role will not be restricted to writing code; it will involve understanding the client's needs and the business landscape and creating a solution that effectively addresses the problems.

This journey begins with comprehending the business's unique requirements. Whether it's developing a rough prototype to validate a concept or building a robust, scalable product, the developer must be flexible, adapting to the context of the project. A software developer's value is also based on their ability to adapt to changing business requirements. When creating a proof of concept, the emphasis might be on speed and functionality rather than optimizing for scale or robustness. It's a balancing act - ensuring the code is effective enough to represent the concept without investing undue time in making it flawless.

On the other end of the spectrum, the developer's focus shifts when building a software product for millions of users. Here, the value is in the ability to craft reliable, efficient, and scalable code. It's about foreseeing potential security issues and performance bottlenecks and designing the architecture to handle significant loads. This ability to balance between conflicting demands is a valuable asset. Their real value lies in the developer's understanding of the spectrum of choices - from prototyping to scalability, from feature prioritization to security considerations - and their ability to make suitable decisions based on the context. Recognizing this is the key to remaining relevant and valuable in their field.

The focus shifts from "what can be built" to "what should be built" and "how it should be built."

As the code generation aspect of a developer's role gets automated, the developers who can seamlessly transition into roles emphasizing strategizing, problem-solving, decision-making, and value creation will be in high demand.

While AI tools are set to become powerful allies in code generation, they are tools. They cannot completely usurp the value developers create; instead, they enhance their capacity to create more value. They would become the "superhumans" in software development, with AI as their sidekick. Conversely, those who resist this change or fail to adapt may struggle. In the age of AI, adaptability is not just a valuable trait but a necessity for survival.

> **'Value Creation' transcends beyond the realm of economics. It's about making a difference, leaving a positive impact, and enhancing the world around us. Value creation is a pursuit not of wealth but of worth. It's about identifying and satisfying needs, finding gaps and bridging them, uncovering opportunities, and seizing them.**

Becoming Superhuman with AI Tools

Having understood the principles of "Original thinking," "Value and Value Creation," let's look at various fields in the real world where intrinsic value can be maximized by effectively utilizing these new-age AI tools, some of which are available today and others we will see soon.

A copywriter's role has traditionally been to craft compelling content that communicates a message effectively and persuasively, whether for an advertisement, a blog, a product description, or any form of content. The value lies not in merely putting words together but creating impactful narratives that resonate with the audience and drive a desired action.

Now, imagine introducing AI into this landscape. AI-powered tools like language models have the potential to generate text that's grammatically correct and contextually relevant. They can churn out content faster than a human writer, eliminating the time and effort spent on the mechanical aspects of writing. Essentially, they can automate the "word production" part of a copywriter's job.

A copywriter's real value lies in understanding the audience, capturing their emotions, empathizing with their needs, and strategically influencing their behavior through words. This aspect of their work is the essence of value creation in copywriting. In the presence of AI, a copywriter becomes more of a strategist and a storyteller. They guide the AI tool to generate content that aligns with the brand voice, the campaign objectives, and the target audience's psyche. The copywriter makes strategic decisions on the tone, style, and narrative structure. They infuse creativity and emotional depth into the AI-generated content.

The role shifts from merely writing to curating, editing, and strategizing. The copywriter ensures that the message isn't just informative, persuasive, and engaging. They shape narratives that resonate with human experiences, aspirations, and emotions. They ensure the content aligns with broader marketing and brand objectives. These skills add value, turn words into compelling narratives, and make a piece of content more than just a collection of sentences.

For instance, an AI tool can generate a basic draft for an ad campaign. However, the copywriter would fine-tune the message, add emotional hooks, and ensure the content aligns with the brand voice and campaign goals. They would decide where to place the call-to-action, what emotional triggers to use, and how to structure the content for maximum impact.

In the age of AI, copywriters who understand and adapt to this shift will become more valuable. They would become curators of narratives, strategists of persuasion, and architects of impactful communication. Conversely, those focused only on the mechanical aspect of writing might find it challenging to stay relevant.

The concept of value creation and AI tools also extends to artists and graphic designers. AI can also be a powerful assistant in these fields, augmenting the creative process rather than replacing it.

At its heart, an artist or a graphic designer's value lies not in the physical act of painting or designing but in their ability to conceive and communicate a vision. It's about conveying a message, evoking an emotion, or narrating a story through visuals. It's about using aesthetics to make an impact and engaging viewers in a way that transcends the visual to the emotional or intellectual. Introduce AI into this equation, and things get interesting. AI tools can now generate artwork, design logos, or even create graphic designs at an astonishing pace. These AI tools can automate the mechanics of drawing, coloring, or creating graphics, performing tasks that may have been time-consuming for a human designer.

However, just as with developers and copywriters, the introduction of AI doesn't diminish the value of the artist or designer. Instead, it augments it. It takes care of the more mundane aspects of the work, leaving artists and designers with more time and mental space to focus on the core aspects of their craft - creativity, conceptualization, and strategy. They use their understanding of color theory, composition, symbolism, cultural contexts, and audience preferences to shape the AI's output, ensuring the result looks good and communicates effectively.

An AI can generate a painting, but the artist commands this AI tool on "what" it should generate. The artist's value isn't in creating those compositions manually but in figuring out the one that best communicates the desired message, tweaking it to perfectly capture the essence of the idea, and applying their unique style to it. In the age of AI, the artist or graphic designer's role evolves from a maker to a curator, a creative director, and a storyteller. They will use AI tools to augment their creativity, not replace it.

Like software developers and copywriters, artists and graphic designers who understand this shift and adapt accordingly will thrive in the age of AI, the successful artist or graphic designer of the future will become a "superhuman" creator, using AI tools to bring their vision to life more efficiently and effectively.

The advent of AI is like a magical wand, bestowing us with previously unfathomable superpowers. It's a revolution set to amplify the value we bring to the table in ways we're just beginning to comprehend.

Imagine, for a moment, that you're a communication strategist. In a traditional setup, your expertise would be confined to your domain - developing communication strategies, planning PR campaigns, or crafting compelling messages. You would rely on teams of copywriters, designers, web developers, and social media managers to execute your vision.

Enter AI, and the game changes dramatically. Equipped with sophisticated AI tools, you're not just a strategist but an all-in-one communication powerhouse. You can navigate seamlessly between creating persuasive copy, designing eye-catching visuals, managing dynamic social media platforms, and even tinkering with web design. All of this without needing to be an expert in coding, design, or copywriting.

Your understanding of the core problem, your strategic thinking, and your creativity are now amplified by AI. AI tools are like your loyal sidekicks, ready to execute your vision, adapt to your strategies, and learn from your feedback. They help you translate your ideas into words, visuals, and code, extending your capabilities beyond traditional boundaries.

For your clients, this transformation is a goldmine. They can now access a spectrum of services from a single source, ensuring consistency in strategy, tone, and aesthetics across different platforms. They save valuable time and resources that would have otherwise been spent on coordinating between multiple teams or agencies.

The result? A higher quality of service, a more significant impact, a greater return on investment for your clients, and a bigger leap forward for you. You're no longer just a communication strategist but a full-fledged digital communication expert.

And this is just one example. The same principle applies to any profession or industry. By understanding the core nature of the problem and leveraging AI, we can all elevate our value creation abilities, expand our range of services, and become more valuable to our clients or employers.

In the realm of translation, the implications of AI are particularly profound and transformative. It's an excellent demonstration of how AI can augment human capabilities and redefine the role of a professional.

Let's envision you as a translator. Traditionally, your work involves converting text from one language to another, ensuring accuracy, and maintaining the original tone and style. This process can be time-consuming, often leaving you with a limited capacity for taking on new projects.

Now, with AI translation tools, the landscape alters drastically. These tools can translate entire documents in a fraction of the time it would take a human, freeing translators to focus on more intricate aspects of the process.

The opportunity here is two-fold: on the one hand, AI amplifies your efficiency, enabling you to handle larger volumes of work with faster turnarounds. On the other hand, it allows you to carve out time to expand your skills, perhaps mastering new languages, diving into specialized fields like legal, technical, or medical translation, or even exploring related areas like localization or international marketing.

Picture this: a client comes to you with a technical manual that needs to be translated into four languages. Without AI, this task could take weeks or even months. But armed with your AI tool, you quickly generate the first draft of translations. You then refine these drafts, ensuring the technical terminologies are accurately translated, the tone is consistent, and cultural nuances are respected. A month-long project could have been completed in a week without compromising quality.

Imagine another scenario: a new client needs a translator well-versed in legal jargon for multiple languages. Thanks to the time saved by AI, you've built expertise in legal translation. You can confidently take on this project, providing high-quality, nuanced translations that a less versatile translator or a standalone AI tool could not.

In essence, you're now offering a package of speed, precision, versatility, and a human touch - a combination that's hard to resist for clients. You've moved the value chain from a translator to a multilingual, multidomain communication specialist.

The transition into this new role might be challenging. It involves learning to trust AI, understanding its strengths and limitations, and honing your skills for higher-level tasks. But the rewards - increased productivity, expanded capabilities, and elevated professional standing - make the journey entirely worthwhile.

Let's explore the fascinating world of law and data analysis as examples of how AI can unlock new horizons.

Picture a paralegal, typically overloaded with document review, contract analysis, and legal research tasks. Paralegals often spend countless hours sifting through pages of legal texts, seeking relevant case laws, or poring over contracts for compliance issues. AI-powered tools can transform this landscape dramatically. These tools can perform tasks such as document review and legal research in a fraction of the time, thanks to advanced Natural Language Processing (NLP) capabilities.

This doesn't mean paralegals will become obsolete. Instead, their role is set to evolve. They can supervise AI tools, ensuring the accuracy of the findings and adding a layer of human analysis that

AI may miss. This free time allows paralegals to develop legal strategy or client management skills. They could even find the time to delve deeper into specialized areas of law. Armed with AI, a paralegal could transform from an auxiliary legal support role to an invaluable strategic asset in a legal team.

Similarly, let's consider a data analyst. Traditionally, a data analyst spends significant time cleaning and organizing data before it's ready for analysis. However, AI-powered tools can automate much of this process, enabling the analyst to focus more on interpreting and drawing insights from the data. This doesn't devalue data analysts; instead, it allows them to focus on the higher-value aspects of their job. They can specialize in interpreting complex data patterns, forecasting trends, or crafting data-driven business strategies. They can spend more time communicating their findings to decision-makers, translating the language of data into actionable business insights.

Yes, mastering new AI tools will be a challenge. Seeing a machine perform tasks you once handled may even be unsettling. However, as these examples show, AI is not here to replace us but to help us ascend the value chain. The key is understanding that the core of our value lies not in the tasks we perform but in the unique human touch we bring to these tasks. As we navigate the AI revolution, let's remember we are not just task performers but problem solvers, strategists, creators, and communicators. We are value creators, and with AI by our side, our potential is limitless.

NOTE OF CAUTION

Our exploration of the future, seen through the lens of the quadrants, has thus far focused heavily on the intersection of human and machine skills. Yet, it's important to remember that this exploration isn't solely about identifying the territories where AI and

THRIVING AMIDST DISRUPTION

automation will reign. It's equally about charting our paths, understanding our unique skills and inclinations, and finding the roles where these intersect with the world's needs.

To future-proof ourselves isn't necessarily to climb the ladder of sophistication to match the pace of AI. It can also mean leaping into other quadrants where our skills resonate more deeply with us, where our work is not just a means of earning a living but also a source of purpose and fulfillment. This is about identifying our "innate skills," the capabilities that come naturally to us and form our essence.

Indeed, the dichotomy of human vs machine skills forces us into introspection, into a journey inward to discover what truly drives us. Are we original thinkers, bursting with creativity and novel ideas? Do we excel at creating value, whether it's through the goods we produce or the services we offer? Wherever these skills lie, there's a space within the quadrants, a niche where we can thrive.

For example, the farming profession lies in Quadrant 3. It might seem an unlikely candidate for future relevance. However, consider a future marked by soaring temperatures and water scarcity, where the ability to grow food efficiently becomes desirable and vital. In this world, the value of an efficient farmer skyrockets.

The farmer who can coax life from the parched earth and utilize every precious drop of water to its maximum effect becomes a linchpin in the future economy. This is a skill that, while not requiring the sophisticated cognitive capabilities associated with AI, requires a deep understanding of the land, a respect for nature's rhythms, and an ability to adapt to changing conditions.

This is not to dismiss the importance of AI and automation. These will undeniably play a significant role in shaping the future. But it is to say that our value as individuals, workers, and contributors to society isn't solely defined by our ability to compete with machines. It's also defined by our capacity to deliver real value and leverage our unique skills and abilities to make a tangible difference in the world.

Supercharged Professions of the Future

Let us continue to explore how various other professions can be augmented with the power of AI in the future. Combining AI and politics can transform the democratic process in ways we could only have imagined a few years ago. By harnessing the power of AI, individuals or small parties can effectively compete with larger, more established ones, thus leveling the playing field.

Imagine you are a politician. Your core value lies in understanding your constituents' needs, crafting policies to address those needs, and communicating effectively with your constituents. Imagine you're an aspiring politician with a small team and limited resources. Traditionally, running an effective campaign would require a vast team to handle various aspects like voter engagement, issue identification, media management, and donation tracking. But, with AI, many of these tasks can be automated or greatly optimized, allowing you to punch well above your weight class.

To begin with, AI-powered chatbots and data analytics could help you understand the key issues concerning your constituents. Through analyzing social media activity, online forums, and other digital platforms, AI could paint a comprehensive picture of the issues people care about the most, their sentiments about various policies, and their overall satisfaction with the state of affairs. This data, gathered in real-time and on a large scale, could be more accurate and insightful than traditional surveys or opinion polls. You wouldn't need an army of volunteers conducting door-to-door surveys. Instead, your AI tool could analyze online data, identify trending issues, and predict voter sentiment based on public opinion on social media and other platforms.

Next, think of personalized voter engagement and constituent communications. Instead of generic messages, AI could help craft personalized communication tailored to individual constituents based on their demographics, political attitudes, and specific concerns. AI can even be used for image and speech recognition to analyze your competitors' speeches and campaign strategies, giving you a competitive edge.

Of course, as a politician, you wouldn't be coding AI models yourself. Instead, you would direct a team of data scientists and AI specialists who could build and implement these systems for you. Your value would lie in understanding the potential of AI and directing its use toward creating value for your constituents.

It's worth noting that while AI can streamline your campaign and enhance your reach, the crux of politics remains human. AI can help identify the issues, but it's up to you to address them with empathy and understanding. AI can help engage voters, but your actions, sincerity, and vision will win their trust. The key is to strike the right balance, leveraging AI to amplify your capabilities without losing the human touch that is the essence of politics. All this doesn't mean AI will make the role of a politician any less demanding or complex. It simply means that AI can be a powerful tool in a politician's arsenal.

As AI becomes more advanced and accessible, it holds the promise of democratizing politics. This new era of politics could see a wave of grassroots candidates with big ideas but limited resources making their mark. With such potential on the horizon, the fusion of AI and politics is exciting.

The future of politics is not just about politicians with the deepest pockets but those with the biggest hearts, resonating visions, and the appropriate AI tools. However, especially as politicians, they must ensure transparency and respect for privacy. They must use AI ethically and responsibly, always putting the interests and rights of their constituents first.

Let's reimagine **architecture and urban design**, where AI and human creativity fusion promises to reshape our built environment.

The core value of an architect lies in their ability to conceive, design, and oversee the creation of functional, safe, aesthetically pleasing, and sustainable spaces. Architects transform visions into concrete realities, creating structures that inspire and facilitate human activity. Imagine an architect's drafting board. Traditionally, it's filled with sketches, plans, blueprints, and material samples. Now, picture an AI-powered drafting board capable of exploring countless design possibilities, optimizing for various factors like sunlight, wind, thermal comfort, or even the cultural context of the space.

Given a boundary of requirements and the essential features of an Architect, AI could sift through a myriad of architectural designs from across the world and throughout history, suggesting unique combinations and inspiring novel architectural ideas. For instance, AI could generate a design based on a given brief and then adjust the design in real time as the architect modifies the brief. The architect might add more sunlight here, more open space there, or change the style from modern to gothic. Each change would trigger new design possibilities, allowing the architect to quickly explore a vast design space.

In urban design, AI could play an even bigger role. The briefs and prompts for the AI systems could include the design of neighborhoods based on the characteristics. It could analyze demographic data, traffic patterns, and environmental conditions to help design urban spaces that are more livable, sustainable, and responsive to citizens' needs.

To grasp the potential here, let us transport ourselves into the future to witness the conversation below:

Urban Designer (UD): Hey AI, I'd like to design a new neighborhood in the northern sector of Cosmocity. Can you assist?

AI: Of course! Let's start with the basics. How large is the area, and what is the primary goal of this neighborhood?

UD: The area is around 300 acres. I envision a sustainable, green neighborhood with residential, commercial, and recreational spaces. It should be pedestrian-friendly and promote community interactions.

AI: Got it. Based on your requirements and the latest urban planning trends, I suggest:

- 60% residential spaces with a mix of low-rise apartments and townhouses.

- 20% commercial areas that include local businesses, cafes, and co-working spaces.

- 15% green recreational spaces with parks, community gardens, and plazas.

- 5% dedicated to public utilities and transportation hubs.

Would you like a preliminary layout?

UD: That sounds good, but let's increase green spaces to 20% and reduce commercial to 15%. I'd also like the residential areas to have rooftop gardens and ensure there's a cycling lane throughout the neighborhood.

AI: Noted. I'll adjust the layout accordingly. Here's the revised preliminary design. [Displays 3D model of neighborhood layout]

UD: This is a good starting point. Can we now focus on the central plaza? I'd like it to be a community gathering hub with space for outdoor events and a children's play area.

AI: Sure! I recommend a multi-level plaza with the following:

- An open amphitheater for events.

- Shaded seating areas using sustainable materials.

- A modular play area for children with interactive installations.

- Water features to create a calming ambiance.

Would you like to see this design?

UD: Yes, please. Also, some smart benches should be integrated with device solar charging stations.

AI: Done! Here's the design for the central plaza with the features you requested. [Displays 3D model of the central plaza.]

The dawn of the AI era holds incredible promise, especially for entrepreneurs. AI provides powerful tools to turn their visionary ideas into reality, even on a bootstrap budget. Entrepreneurs are the problem solvers and innovators of the business world. These pioneers, always quick to spot gaps and forge innovative paths, now have AI as their ultimate sidekick. Imagine a world where even the smallest entrepreneurial unit, fueled by groundbreaking ideas, can rival the impact of vast organizations, all thanks to AI's might.

Think of AI as your ultimate Swiss Army knife in the startup battlefield. It's the secret sauce that can expedite product development, morphing your embryonic ideas into tangible prototypes and even full-blown products at warp speed. Mundane tasks that once gobbled up your time—scheduling, email sorting, accounting—can be effortlessly outsourced to AI minions, liberating you to architect grand strategies and forge invaluable partnerships.

But the magic of AI extends even to the financial fortresses of fundraising. Picture AI as your tireless emissary, sifting through oceans of data to pinpoint potential investors, crafting resonating pitch decks, and managing investor relations with automated finesse. It's like having a Wall Street whiz, a Madison Avenue creative, and a Silicon Valley engineer rolled into one—available at your fingertips.

Yet, the crowning jewel of AI's potential is its scalability. Entrepreneurs can now dream big—big. Starting solo and scaling to serve millions becomes a possibility and a palpable reality with AI.

It's critical, however, not to mistake this for a robotic coup of the entrepreneurial spirit. No algorithm can replicate the uniquely human sparks of vision, creativity, and leadership. The entrepreneur evolves from a creator into a curator in the AI-driven world. This visionary choreographs a mesmerizing dance between human ingenuity and machine precision.

So, whether you're a fledgling founder or a veteran business virtuoso, know this: AI isn't your competition; it's your catapult. It's the wind beneath your wings, ready to propel your entrepreneurial dreams into the stratosphere of reality.

On the other side of the spectrum, imagine a future where you're a passionate home cook, stirring up magic with your pots and pans. With a small setup in your kitchen and a vision to create delicious meals that bring joy to people's lives, you may wonder how AI can assist you.

Buckle up because we'll add digital spice to your culinary journey!

At the core, a chef's value lies in conceiving delectable dishes and executing those ideas precisely. This requires a blend of creativity, an understanding of flavors, and a deft hand at preparation. Add a dash of AI into this mix, and your possibilities for innovation and scaling up suddenly multiply.

Consider AI as your digital sous-chef, which can aid in various aspects of your culinary business. For instance, imagine having an AI tool that can scan through countless recipes worldwide, finding unique combinations of ingredients and cooking techniques that might inspire your next dish.

AI could help you discover hidden gems - the proverbial "secret sauce" - in global cuisine, allowing you to introduce a unique flavor profile that sets your food apart. AI can even assist you in adjusting portions or ingredients, enabling you to easily cater to dietary restrictions or preferences.

With AI by your side, the dream of becoming a neighborhood sensation or building a food delivery empire becomes more attainable. And the best part? To do this, you don't have to be a culinary school graduate or tech wizard. Suppose you have a knack for cooking and a willingness to experiment with AI. In that case, you can start your gastronomic adventure from your kitchen.

AI can be a great tool for everyone, from computer coders to writers and politicians to home cooks. It helps us do more faster and in new ways. Like a super-powered sidekick, AI can help us shine in our jobs by doing the heavy lifting and letting us focus on the most important parts.

But as we enjoy these benefits, we also need to be careful. AI is a powerful tool; like all powerful tools, it can be used incorrectly. We mustn't use AI to harm others or steal their creative ideas. We also need to ensure that AI doesn't create new unfairness or make things more unequal.

So, as we step into this exciting future with AI, we need to be both excited and careful. We are shaping this future, and we can make it great. We can use AI to make our jobs more fulfilling and make the world fairer. But we need to do it thoughtfully, responsibly, and together. AI is a tool that should help us, not overpower us. In the upcoming chapter, let's look at the roles of other drivers of society: "Enterprises," "Investors," and "Governments" to drive this transformation responsibly.

6

ABSORBING THE DISRUPTION: ENTERPRISES

Why should a successful business leader, perhaps one steering a ship far from the shores of technology, concern themselves with developments in AI? The answer is surprisingly simple: no enterprise, no matter its size, domain, or geographic location, is immune to the ripples of AI's growing influence. The digital ecosystem we inhabit and upon which our enterprises depend is morphing under the weight of AI's potential.

Think back to the advent of the internet. Initially, it was merely a medium for exchanging emails and sharing files. The connection between this new technology and a travel company, a retail business, or an engineering firm might have seemed tenuous at best. Yet, fast-forward a few decades, no company today operates without the internet's pervasive influence.

Businesses that once thrived solely on physical interactions underwent a radical digital transformation. Simultaneously, entirely new industries emerged — social media, cybersecurity, digital marketing, to name a few. The internet became the fulcrum on which businesses pivoted, and those leaders who had the foresight to anticipate this and could imagine harnessing its latent power emerged as the winners of this seismic shift. Now, AI stands on the brink of causing a similar upheaval. Its potential is vast, arguably more so than the internet, and its implications are far-reaching. It's not about immediate applicability but about preparing for the inevitable evolution.

The digital landscape is a vast, interconnected web, and AI is poised to become its beating heart. Its tendrils will reach out, touching and transforming every aspect of business. It promises a level of automation, accuracy, and insight unprecedented in human history. Yet, AI isn't just about creating smarter processes; it's about opening up new vistas of opportunity.

It can help you understand your customers better, open up new revenue streams, create new products and services, and even venture into uncharted business territories. Ignoring AI could equate to turning a blind eye to opportunities, efficiencies, and insights that could potentially revolutionize their business. Most importantly, not understanding AI could yield a competitive advantage to those who do.

An in-depth understanding of AI is not a luxury for business leaders but an imperative. It doesn't mean every CEO and executive needs to become an AI expert. Still, they should be able to appreciate AI's capabilities and limitations, envision its potential impact, and chart a strategic course for AI adoption and integration in their business. The goal is not to fear AI's touch but to embrace it, understanding that it will be the new lifeblood of innovation, productivity, and competitive differentiation in this digital landscape.

We are witnessing a technological transformation surging forward at an unparalleled pace. Once we were in awe of digital computing, it took 40 years to become the backbone of our societies. Then came the internet and cloud computing, which shrunk that timeline to a mere 20 years and placed the world at our fingertips. Now, we are grappling with AI, a technology that promises to become ingrained in every fabric of our lives in an astounding 5-10 years. As with every major shift in human history, it's a time of profound excitement yet daunting challenges. The pervasive integration of AI into our economies also illuminates fears of workforce displacement, ethical quandaries, and questions about its potential implications for social and economic equity. Every business leader needs to care about AI because, ready or not, AI is here and poised to reshape the business world as we know it.

AI Literacy - The New Language of Leadership

Being at the helm of a corporation has always been a Herculean task. In leading enterprises, especially in today's volatile, uncertain, complex, and ambiguous world, executives have their work cut out for them. They shoulder the hefty responsibility of guiding their teams and the livelihoods of all under their purview. Theirs is a life in constant flux that involves staying ahead in a fiercely competitive landscape, strategizing years into the future, ensuring a steady flow of capital, and generating value to satiate the appetites of investors and shareholders.

Amid these formidable responsibilities, their minds are ceaselessly churning — at work, at the dinner table, during holidays, in the dead of the night. The quest for value creation, a relentless pursuit of creativity, innovation, and efficiency, engulfs their existence. A tough life, you might think, and you wouldn't be wrong.

Yet, here enters the latest curveball: AI, specifically Generative AI, the new entrant and game-changer in the competitive playground. This potent, transformative technology promises the sun, moon, and stars yet asks for the one thing most in short supply - mental bandwidth. How do leaders, already stretched to their limits, find the time and space to comprehend AI's real opportunities and applicability while keeping the wheels of their enterprise turning?

This, dear reader, is where AI literacy for enterprises comes into play. It is the bridge that connects the technical realm of AI to the practical business world, a lens through which executives can glimpse the future, make sense of it, and begin to mold it to their advantage. It is about comprehending AI not as an abstract concept but as a tool that, when correctly harnessed, can unlock unprecedented value across the organization.

We will delve into this new language of leadership, exploring the facets of AI literacy necessary for executives and corporations. The focus isn't just on understanding how AI works but also its practical applications, ethical implications, and, most importantly, its potential to deliver a competitive edge. It is about decoding this complex technology to wield it not as just a sword to defeat competitors but as a compass to navigate the uncharted territories of the business landscape.

As we embark on this exploration, remember that the goal isn't to become an AI expert but rather an AI-literate leader - one who knows enough to ask the right questions, grasp the potential, understand the risks, and guide their teams toward a future augmented, not dominated, by AI.

Visualize this: it's a brisk Monday morning in the executive suite. The aroma of freshly brewed coffee wafts through the air as a team leader or a CEO, invigorated by a recent AI podcast, declares with contagious enthusiasm, "It's high time we harness the power of AI! The future is knocking at our door, and we must answer with innovation!"

It's a scenario played out more often than you'd think. Having caught the winds of the AI hype, CEOs and executives stand before their teams and announce, "We must leverage Generative AI. Let's get those AI tools going. Where's our AI migration strategy?" The response is often a room full of bemused faces and the unspoken question, "So, where do we start?" They've heard this kind of enthusiasm before, but understanding how to translate it into actionable plans remains elusive. Does that ring a bell? If so, you're not alone. This scene, often fodder for workplace humor, reflects the reality in numerous enterprises. The whirlwinds of AI hype sweep executives off their feet, and in an attempt to stay afloat, the buzzword 'AI' becomes a Monday morning mantra.

Why does this gulf exist between executive enthusiasm and operational reality? The answer lies in the endless storm of AI chatter that fills industry publications and fuels executive imaginations. As we progress, this book will serve as your toolkit to cut through the clamor of AI hype. Before we dive in, these strategies are universally applicable and domain-agnostic. Whether you're a tech tycoon, culinary giant, healthcare pioneer, or financial mogul, these frameworks will serve as your north star, guiding you toward AI clarity.

Remember, embracing AI isn't about jumping onto a trend bandwagon. It's about aligning AI with your enterprise's vision, values, and goals.

Keeping away from enterprise jargon, we aim to achieve clarity by answering three main questions. Three reasons "why" should enterprises adopt AI? Four channels of "What" can you do with AI? Three strategies on "How" to step into the future. The answers give a simple yet comprehensive and powerful framework for an industry-agnostic AI adoption pipeline.

Firstly, we'll tackle the gnarling question: "What challenges does AI pose to enterprises?" AI proliferation brings unique challenges like learning a foreign language or mastering baking sourdough during a global pandemic. Secondly, we'll address the million-dollar question: "How do we find the real value of AI applicability?" Here, we'll talk about how AI isn't a magic wand (though it does have a knack for making your competitors disappear if wielded right). AI's value lies not in its novelty but in its ability to solve business problems, streamline operations, and create unique customer experiences. Our frameworks will help you uncover these golden nuggets of opportunity beneath the AI gold rush. Thirdly, let's examine the importance of "Data" and strategies for leveraging AI for exponential growth without disrupting your workforce.

Adapting to AI is like learning to dance with a new partner—one whose steps are powered by algorithms and predictive analysis. Balancing your human workforce's strengths with AI's efficiency is a delicate tango.

AI LITERACY CHEATSHEET FOR ENTERPRISES

WHY CARE ABOUT AI

1. Sword of external Disruption with value proposition

2. Competitive Edge of Operational Excellence

3. Talent magnet - Beacon of attraction

WHAT DO TO WITH AI

1. Streamline Internal process

2. Accelerated product Development

3. AI integration in products

4. New AI based products

HOW TO TRANSITION WITH AI

1. Data - data - Data

2. Handling Workforce reorganizatioin

3. Recalibrating Frontier and transformative activties

AI Adoption Imperative

The Sword of Disruption

"Disruption". This term, once associated with unruly behavior, has taken on a radically different meaning in the business world. Today, it represents a shake-up of established norms and a revolution of conventional practices, and AI is at the forefront of this change.

The word 'disruption' is often tied to worker job security in the age of AI, but its ramifications stretch far beyond. As a potent force, it can destabilize even the most established businesses. Let's dissect this further, starting with the first core aspect of "business value proposition."

Imagine being at the helm of a market-leading company, enjoying years of success and customer loyalty. You're not just surviving; you're thriving. You've cornered a significant market share, and your name is synonymous with your industry. You're sitting pretty, right? Yet, on the horizon, a storm is brewing. A fledgling startup, built from the ground up with AI woven into its DNA, is gaining momentum. Their value proposition is different. It's innovative. It's AI-driven.

In the world of AI, it's no longer just about who has been around the longest or who has the deepest pockets. Today, a fledgling startup, built from the ground up around AI — an "AI Native" — could threaten your position. Such AI-native enterprises can offer superior products or services, personalized experiences, and unrivaled efficiency, all at competitive prices. Boosted by "Investor confidence" in disrupting an existing market, these companies will leverage AI to deliver superior value. These newcomers can offer better, more tailored products and services. Over time, they could start nibbling at the edges of your customer base, steadily eroding your market share and potentially threatening your survival. The speed of this evolution can be breathtaking, and the incumbents are left flat-footed.

Consider the insurance sector. Traditional insurance companies have long thrived on offering standard insurance plans. But what if a new, AI-driven competitor enters the market, offering customized insurance plans tailored to each customer's needs, risks, and lifestyle? Their value proposition is irresistible, and they could quickly seize a considerable market share.

Or, look at pharmaceutical companies. Drug discovery is a long and costly process. However, an AI-native firm has entered the scene, using machine learning to accelerate drug discovery and reduce costs dramatically. Suddenly, they're producing effective medications faster and cheaper than established pharmaceutical giants. In the legal domain, traditional firms may charge hefty fees for services that an AI-driven competitor can provide more swiftly and cost-effectively. AI can analyze thousands of legal documents in a fraction of the time it would take a human, enabling these firms to offer quick and affordable legal services.

The list of examples could go on. A publishing house that employs AI to tailor content to readers' preferences and reading habits or deliver news tailored to every individual viewer. And let's not forget entirely new businesses that we can scarcely imagine today. AI could lead to entirely new industries and services, just like the internet did with social media and e-commerce.

The key takeaway is that no industry is immune to AI's disruptive potential. Incumbents must stay vigilant, constantly reevaluating their value proposition in light of emerging AI capabilities. They need to understand AI's potential in their industry, identify potential threats and opportunities, and be prepared to pivot or evolve their business model. Ignoring AI's potential is no longer an option; it's a ticket to obsolescence. AI disruption is real and happening faster than most can anticipate.

The Internal Revolution

The second form of disruption resulting from AI isn't about novel, AI-driven competitors. Instead, it involves a transformation from within. This second disruptive facet of AI lies in its potential to optimize internal business processes. When AI is used to enhance operational efficiency, businesses can deliver the same product or service but with improved quality, faster, or even at a reduced cost. This, too, can challenge the market status quo, leading to a shift in market share.

Take an insurance firm, for instance. It has hundreds of

thousands of policy documents and client data. Previously, a customer support agent would have to manually search through this massive data pool to address a customer query. This process could be time-consuming and prone to errors. However, the company can drastically improve this process with AI. Agents can get precise, real-time information at their fingertips by implementing a semantic search system powered by AI. This speeds up customer service and improves quality, enhancing customer satisfaction.

Imagine two manufacturing companies competing in the same market. They have a similar product lineup and have been neck-and-neck in market share for years. However, one company invests in AI to optimize its production processes. Machine learning algorithms predict maintenance needs, minimize downtime, and maximize efficiency. AI-driven analytics provide real-time insights into supply chain inefficiencies, allowing the company to reduce waste and improve sourcing.

The result? This company can now produce the same goods more quickly, at a lower cost, and potentially even with better quality. It passes these savings and benefits onto consumers, giving it a competitive advantage. Over time, it starts to carve a larger market share pie.

These examples underline that AI disruption is not solely an external threat posed by new, AI-native startups. It also stems from within an industry, as existing competitors leverage AI to optimize their operations, offer better customer experiences, and innovate faster.

To ignore this aspect of AI disruption is to risk being left behind in the race, unable to match the efficiency, quality, and innovation offered by AI-empowered competitors. Understanding and embracing AI is not just about future-proofing your business - it's about securing a competitive advantage today.

The Talent Magnet: AI as a Beacon of Attraction

The third key aspect of AI's influence on businesses lies in human resources. As the saying goes, 'a company is only as good as its people.' Quality manpower can be difficult to find, and in a rapidly evolving technological landscape, businesses that fail to adapt may lose the race for talent.

In the AI era, jobs involving lower-level cognitive tasks will increasingly be performed by machines, a shift we've already started to witness. This isn't necessarily a cause for alarm; it opens up opportunities for humans to engage in more complex, higher-value tasks. The displacement effect of AI isn't a job eliminator but a job evolver. Employees who can effectively use AI tools, understand their nuances, and leverage their strengths for complex problem-solving will become increasingly valuable.

Imagine an engineer or a data scientist choosing between two firms. One is a traditional company, still relying largely on manual processes and legacy systems. The other is an AI-forward company investing in the latest AI tools, promoting a culture of continual learning, and pushing the boundaries of what's possible with AI. The latter is likely a much more attractive proposition for talent eager to work on the cutting edge of their field.

As such, companies that embrace and utilize cutting-edge AI technologies can become talent magnets. Being at the forefront of AI adoption sends a clear message to current and potential employees: the company is forward-thinking, innovative, and committed to empowering its workforce with the best tools available. This can incentivize talented, capable employees to stay, nurturing their growth and encouraging them to reach new heights of productivity.

By automating routine tasks, AI allows employees to focus on more challenging, rewarding work, leading to higher job satisfaction.

Employees can spend their time on creative problem-solving, strategic thinking, and innovation, all of which contribute to a more engaging and fulfilling work environment, resulting in a win-win for both sides. The company's AI-forward stance could attract high-caliber talent from the outside. Talented professionals often seek workplaces that offer opportunities for learning and growth. They aspire to work with cutting-edge tools and technologies, and AI is certainly at the top of that list.

Conversely, failing to integrate AI into your business practices could lead to a talent drain. Employees, especially the tech-savvy ones, may feel stagnated and seek greener pastures, potentially joining your competitors. Furthermore, your company may struggle to attract new talent. Aspiring professionals, especially tech-related professionals, may perceive your company as out-of-touch or stagnant, hindering recruitment efforts.

In attracting and retaining talent, it's crucial to note that investing in AI is not just about purchasing and deploying the latest AI tools. It's also about investing in the training and development of employees, enabling them to harness these tools effectively. This commitment to employee growth further enhances a company's appeal to prospective employees and helps retain existing talent. Your employees are your most valuable asset, and AI can empower them to perform at their best.

While it can be daunting to keep track of various advancements in AI technologies spawning every day, it can be comforting to know that there are only four strategic ways in which they can be applied in any enterprise.

Avenues for AI Adoption

Streamlining Internal Processes

The first approach to adopting AI in your organization is improving and streamlining your internal operations. Since our workplaces are more digitized than ever, most of our work involves interaction with text-based data. Businesses are inundated with text-based information, be it emails, reports, meeting notes, or even social media posts. Traditionally deemed 'unstructured, this data can now be effectively managed, understood, and utilized with modern AI tools.

Let's delve into some ways AI can revolutionize various internal processes:

1. **Human Resources**: AI tools can automate repetitive tasks such as scheduling interviews, screening resumes, and onboarding new employees. Natural language processing (NLP) can help analyze job descriptions and match them with suitable resumes, speeding up recruitment. AI can also assist in gauging employee sentiment through internal communications analysis, providing valuable insights to boost employee satisfaction and retention.

2. **Meeting Management**: Generative AI can transcribe meeting notes, ensuring no critical point is missed. AI can further analyze the discussions, highlight important tasks and deadlines, and assign responsibilities, thereby enhancing the productivity of meetings.

3. **Marketing and Sales**: AI tools can aid market research by scouring the internet for data and trends. They can also personalize marketing content to individual customers, improving engagement rates. In sales, AI can analyze customer communication to understand their needs better,

thereby improving customer service and ultimately driving sales.

4. **Customer Support**: AI chatbots can handle customer queries round the clock, providing instant support. NLP can be used to analyze customer feedback and reviews, providing insights to improve products and services.

5. **Project Management**: AI can assist in tracking progress, forecasting project timelines, and identifying potential risks. It can analyze past projects to provide insights into improving efficiency and reducing project costs.

6. **Reporting**: AI can generate and analyze reports in real time, providing insights to make data-driven decisions. It can also predict future trends, enabling proactive business strategies.

7. **Transcending Language Barriers**: In global businesses operating across multiple geographical locations, AI can assist in language translation and localization of content, ensuring effective communication across all regions.

By embracing AI tools in these areas, businesses can improve operational efficiency and free up their workforce for higher cognitive tasks.

The benefits are manifold: quicker response times, reduced costs, increased productivity, and happier employees. As such, adopting AI for internal operations is a strategy that businesses of all types and sizes can benefit from.

Accelerating Product Development

The second dimension of AI adoption involves leveraging AI to develop your products more efficiently, with higher quality and a leaner team. As AI technology advances, an increasing number of tasks involved in product development can be automated or enhanced, allowing teams to focus on higher-level decision-making and creative tasks.

Here are a few ways AI can enhance your product development process:

1. **AI in Software**: AI-powered coding assistants are becoming increasingly sophisticated, with capabilities extending from identifying syntax errors to suggesting optimal code snippets. These tools can save developers time and effort, streamline debugging, and help novices learn to code more efficiently.

2. **Graphics and Design**: AI can generate creative content such as logos, layouts, or UI components based on given specifications, significantly accelerating the design process. Furthermore, AI can facilitate rapid prototyping by rendering 3D models, predicting user interaction, and running design tests.

3. **Content Creation**: AI can assist in generating content, including text for product descriptions, marketing materials, blog posts, and more. Using Natural Language Generation (NLG), AI can create grammatically correct, contextually relevant, customized content to drive engagement and conversions.

4. **Film and Animation**: AI can be used to automate tasks in the post-production process, such as color grading, sound editing, and even creating special effects. In animation, AI can streamline the animation process by automatically generating in-between frames, a task that is labor-intensive when done manually.

5. **Testing and Quality Assurance**: AI can automate the

testing process, which is crucial for product development. Machine learning algorithms can learn from past tests to identify patterns and predict where errors may occur in the future. This allows for more comprehensive and efficient testing, leading to a higher-quality end product.

6. **Future Possibilities**: As AI technology continues to evolve, we can expect even more applications in product development. For instance, generative AI could automate music creation for products that require sound design or help generate plot ideas for game developers.

The potential is vast and continually expanding as AI technology continues to advance. It is not just about getting products out to market faster; it's about leveraging AI to create better products that truly resonate with your target audience.

Integrating AI into existing Products

The revolutionary potential of text-based generative AI lies in streamlining internal processes or improving the speed of product development and in reinventing the very nature of your products and services. Through understanding and generating human language, these AI technologies can bring about a new level of interactivity, personalization, and value creation in your offerings. Let's delve deeper into this through some real-world applications:

1. **E-commerce**: Generative AI can utterly transform online shopping. Instead of navigating through layers of product categories or typing in exact product names, customers can simply express their needs in their own words. For instance, a customer might type, "I need to organize my kitchen better." The AI, comprehending"this query, could suggest a range of relevant products like kitchen organizers, storage containers, and shelf dividers. This AI capability mimics customers' interaction in local stores,

enhancing their shopping experience and creating tremendous value.

2. **Customer Service**: AI chatbots powered by generative AI can handle customer queries in a much more sophisticated and human-like manner. By understanding the context and nuances of customer conversations, these AI chatbots can provide highly personalized and effective solutions, boosting customer satisfaction and loyalty.

3. **Education**: AI can be integrated into EdTech platforms to support personalized learning. For example, AI can understand a student's written question, figure out the underlying concept the student is struggling with, and generate a detailed, step-by-step explanation.

4. **Content Creation**: AI can assist or even automate the creation of various types of content, from marketing copy to news articles. This can make content production faster, more cost-effective, and scalable.

5. **Financial Services**: Banks and financial institutions can use AI to interpret customers' financial needs in natural language and offer personalized advice or product recommendations.

6. **Healthcare**: AI can assist in patient triage by understanding the symptoms described by the patient in natural language and offering preliminary advice or directing them to the appropriate medical professional.

7. **Transportation**: Ride-hailing services can use AI to understand requests expressed in natural language, such as "I need a car that can fit a bicycle" or "I need to get to the airport by 5 PM".

In each of these cases, generative AI enhances the functionality and efficiency of the product or service and elevates the customer experience by addressing their needs more personalized and intuitively. This value creation at the individual customer level can ultimately drive business growth, market differentiation, and long-term success in the era of AI.

Develop AI-Based Solutions and Reinvent Your Business

And here we reach our final lap on this thrilling journey of understanding and embracing AI: becoming an AI-based solution creator. No, we're not suggesting you abandon your core business and declare yourself the next big AI startup. Rather, we're discussing spotting opportunities within your industry or market, leveraging your unique insights, and architecting AI-based solutions that address those opportunities. We're at an exhilarating moment in history where the uncharted territory of AI has started to unravel, shedding its complex layers and becoming more accessible and meaningful for a broad range of industries. Suppose you can discern the promising glimmers of AI applicability within your specific domain. In that case, you can become a spectator of the AI revolution and actively participate in shaping it.

How often these opportunities lurk in your day-to-day operations or customer interactions might surprise you. It's all about viewing your enterprise from the lens of AI and the potential value it can bring. This mirrors the emergence of cloud computing – a technological paradigm initially obscure to many but soon became omnipresent. Amazon's journey with AWS is a perfect example. It started as an internal project to meet their needs but evolved into a behemoth offering catering to a vast and varied clientele. A book-selling platform that morphed into a general e-commerce giant. Amazon then made an audacious leap into a seemingly unrelated area: cloud computing. What started as an internal project, Amazon Web Services (AWS), is now a colossal enterprise in its own right, powering vast swaths of the internet and catapulting Amazon into the tech stratosphere. Why? Because they saw a need, they innovated and weren't afraid to branch out.

Now, let's circle back to where we began, remembering those well-intentioned but often bewildering directives: "Let's use AI!". It's clear now that there's so much more to "using AI" than just adopting the latest trendy tech. You're not just shopping for a new gadget; you're embarking on a transformative journey that requires a solid strategy tailored to your business specifics and AI's unique value. Here, then, is your AI compass:

- Improving internal processes and efficiency,

- Accelerating product development,

- Enhancing products with AI features,

- Creating AI-based solutions to serve your market needs.

Of course, these aren't strict boundaries. They overlap and feed into each other. The trick is identifying areas where AI can add the most value to your business. Could AI streamline your workflows? Could it take your product to a whole new level? Could you be the next Amazon, transforming an internal project into a global powerhouse?

It's a bit like a treasure hunt. Only, in this hunt, you're not merely looking for gold but for ways to transform lead into gold. This is where the magic of AI comes into play, and the fun part? You get to be the alchemist.

Yes, the task of harnessing AI might feel like herding cats at times—especially those quantum, self-replicating, algorithmic cats – but remember: you're not alone, and the opportunity to create incredible value has never been more exciting.

Data, Data, and Data

Today's AI landscape can be compared to the Wild West of the 19th century – untamed, full of promise, and brimming with opportunity. Yet, it is also characterized by the uncertainties and challenges typical of any new frontier. Therefore, navigating this terrain with open eyes and an informed mind is crucial. In this context, 'AI literacy' is not a luxury but a necessity for organizations looking to thrive.

However, the rapid evolution of AI technologies and the myriad of applications that sprout with every passing month can make this task daunting. While the vision of a fully matured, 'plug-and-play' AI solution that seamlessly integrates into your organization's operations is indeed captivating, it remains, at least for now, just that – a vision.

In reality, AI technology is in its adolescence. Its capabilities are growing and expanding incredibly, but it is still far from the 'all-knowing' and 'all-solving' entity that some imagine it to be. New solutions, business processes, and models are being churned every month.

This is not to undermine the remarkable feats that AI has achieved or its potential to transform industries. Rather, it is a call to action for enterprise leaders to take a more hands-on and informed approach in their AI adoption journey. Understanding AI's limitations and growing pains, the need for customization, and the importance of regular updates and maintenance is essential. These are crucial elements to harnessing the benefits of AI technology effectively.

At this juncture, 'AI literacy' should be regarded as a core competency for the IT department and the entire organization. It's akin to digital literacy in the 2000s – integral to understanding the changing dynamics of the industry and essential for survival. Being 'AI literate' empowers businesses to sift through the hype, identify valuable AI solutions relevant to their operations, and, more importantly, adapt as technology evolves.

Just as we don't expect every employee to be a data scientist, 'AI literacy' doesn't necessarily mean understanding the intricacies of neural networks or the math behind machine learning algorithms. It's about developing a fundamental understanding of AI's capabilities and limitations, recognizing its potential impact on your industry, and cultivating the foresight to anticipate its future trajectory. It is imperative for the "Top and Senior level Management" to get a birds-eye view of the capabilities, limitations, strategic opportunities, and challenges. The "mid-level managers and technical-operational team" should be aware of continuous developments in new tools and their applicability in everyday operations. This creates a solid balance for the path ahead together.

In the face of an AI revolution, those who are prepared, aware, and capable of navigating its uncertainties are the ones who will truly seize the opportunities it presents. And therein lies the role of every enterprise leader today – to equip your organization with the knowledge and the tools to architect the AI disruption.

As we continue our journey into the AI-led business world, we encounter the raw material that fuels these advancements: data. In this AI-enhanced ecosystem, data is not just a by-product of your operations or a mere repository of past events—it becomes the lifeblood of your enterprise's potential for innovation and competitive edge. Its importance cannot be overstated.

Data is not just the king in AI—it's the entire kingdom. Quality data is the fundamental pillar on which AI solutions can be built. AI, especially generative AI, thrives on the richness, diversity, and granularity of data at its disposal. It's similar to a gourmet chef. Even with the best culinary skills, the chef cannot do much without the right ingredients. Likewise, even the most advanced AI models are as good as the data they are fed. If the data is unstructured, disorganized, or poor quality, even the most sophisticated AI tools can falter, delivering less than stellar results.

Moreover, as each enterprise has unique data sets gathered from its specific operations and customer interactions, the outcomes from AI tools can vastly differ. Thus, it isn't just about having AI; it's about leveraging it with the right data. That's why an enterprise's strategy towards data management and organization is crucial.

It's important to remember that AI isn't a magical wand you wave to solve all problems—it's more of a high-performance engine. Like any engine, it needs the right fuel—in this case, high-quality data— to run smoothly and at optimal capacity. And just as a car's performance can be impeded by bad fuel, so can an AI's performance be diminished by poor data.

Every piece of data, from the most mundane transaction to the most sensitive customer feedback, has the potential to train your AI tools better and to extract insights that may otherwise remain hidden. The same AI solution can yield different results for different enterprises, and this difference, more often than not, comes from the data quality, its organization, and how it's managed.

So, as we explore the potential of AI, remember that this isn't a race against other enterprises. It's a race against your ability to effectively generate, manage, and use data. The better you are at this, the more beneficial AI will be to your organization.

Step back before integrating AI into your operations or investing heavily in AI tools. Look at your data. Is it well-organized? Is it of high quality? Is it being effectively managed? Because these factors will significantly impact your AI strategy's success. It's a journey, and data is your compass. Handle it well, and it will guide you to the right path.

These insights about the pivotal role of data in AI initiatives form the foundation of a robust enterprise AI strategy. It cannot be overstressed how important it is to understand the power of your data and safeguard it effectively. It's akin to protecting your most valuable treasure.

The era of AI proliferation will bring an influx of AI solution providers. Each will offer a dazzling array of products, services, and promises for an AI-enhanced future. However, understanding how these solutions interact with your data is paramount in this rapid-paced, high-stakes landscape. Your data is the lifeblood of your AI operations—it fuels the AI engines, refines their efficiency, and determines the overall success of your AI venture. Hence, preserving the sanctity of your data ecosystem is a non-negotiable requirement.

So, before you align with an AI solution provider, equip yourself with probing questions that reveal the nuances of how the solution interacts with your data. Here are the critical ones you need to ask:

1. Where is the AI solution operating? Is it cloud-based, on-premises, or a hybrid model? Understanding this will give you insights into data accessibility, transfer, and storage.

2. How does the solution access my data, and how is my data used for training? This is critical to understanding the 'behind-the-scenes' working of the solution and its effect on your data ecosystem.

3. Will my data leave the confines of my ecosystem? If so, how and where? The security and privacy of your data should always be a primary concern.

4. Suppose the core algorithm improves by training on my data. How can I ensure that these improvements don't find their way into software upgrades and get distributed to other companies using your solutions? Here, you're ensuring that

your data-derived benefits don't become a competitive advantage for others.

5. How does the AI solution provider customize algorithms based on my data? Are integration specialists working on my premises to structure and feed the data into their systems? Here, you're ensuring the best use of your data for maximum AI effectiveness.

Remember, the AI marketplace will teem with providers offering you the moon. Still, the most promising prospect for your organization will be the one that treats your data with the respect and discretion it deserves. In the world of AI, your data is your strength. Guard it fiercely and use it wisely; the AI landscape will benefit you.

It's not just about your enterprise's internal data. An equally, if not more critical aspect, is customer data. It's the lifeline of your business—vital for understanding your customers, improving products and services, and making strategic decisions. However, handling customer data is like walking on a tightrope. On the one hand, it's an invaluable asset for your AI initiatives, while on the other, it's a responsibility that demands utmost caution and respect.

Security and privacy of customer data should be at the heart of your AI strategy. Any mishandling, breaches, or unauthorized access to customer data can be catastrophic for your business. It can lead to regulatory penalties, a tarnished reputation, and a severe loss of customer trust—often irreversible consequences. Therefore, ensuring stringent data security measures and privacy controls is not optional; it's a business imperative.

Data security in the age of AI isn't an optional endeavor—it's a business imperative. It's not just about protecting assets but upholding trust.

Moreover, another pressing issue that needs addressing is the ethical use of data beyond just data protection. AI, with its potential to harness large amounts of data for insights and decision-making, can pose significant ethical challenges. In the race to exploit AI, it's essential not to lose sight of moral and ethical boundaries. Fairness, transparency, and respect for privacy must guide all your data-related practices. It's about accountability to customers, employees, stakeholders, and society.

When considering AI solutions, ensure that they align with these principles. Ask providers about their data ethics policies and practices. Understand how they ensure data fairness and prevent bias in their algorithms. Find out how they maintain transparency in their operations. Most importantly, they must verify their commitment to privacy and their measures to protect it.

In the world of AI, data is not a King. But it's a kingdom that requires meticulous governance. It involves safeguarding your most valuable asset—internal and customer data—from threats, using it responsibly, and ensuring your AI solution providers share your commitment to data security, privacy, and ethics. It's not just about reaping the benefits of AI but doing so in a manner that respects your customers' trust and upholds your business's integrity.

Tackling Challenges in AI Transition

Transitioning to AI isn't just a matter of switching out tools or systems—it's a transformation that extends to the organization's heart, touching everything from technology to culture, operations to employee roles. Just as a finely tuned orchestra requires carefully balancing instruments, so does the symphony of AI implementation. The task is as delicate as it is daunting, and therefore, it should be approached with precision, understanding, and, most importantly, a well-thought-out plan.

Consider, if you will, the technological aspects. AI systems are not simple plug-and-play solutions; they are complex networks that require comprehensive integration with existing systems. Merely procuring an AI solution isn't enough. The real challenge is understanding how it intertwines with your organization's infrastructure, impacts your data flow, and coexists with other systems. The crux here isn't acquisition but assimilation.

But this is just one side of the coin. The flip side—often overlooked but equally critical—is the cultural shift. Introducing AI into your workspace isn't merely about introducing a new technology but a new way of working. It's about embracing a change that might disrupt established routines, alter long-standing processes, and redefine employee roles. The organization needs to be agile, the employees open-minded and adaptable, the leadership strong, visionary, and, most importantly, understanding.

And herein lies the risk: An unsuccessful AI implementation isn't just about failed technology—it's about wasted resources, squandered time, and eroded morale. It's akin to buying a high-performance sports car but not knowing how to drive it—or worse, driving it down the wrong road.

Approach AI implementation with caution but also with excitement. Plan meticulously. Understand deeply. Train rigorously. Communicate openly. Implement gradually. These are steps towards not just a successful AI implementation but also a successful AI future. Remember, AI is more than a tool—it's a transformation. And like all great transformations, it deserves to be handled with care, strategy, and a dash of audacity.

Workforce reorganization—the "AI-driven metamorphosis" in human resource management. It's the specter at the feast, the question that hangs in the air, the topic that sends a collective shiver down the spine of any corporate gathering: What happens to our employees in the age of AI?

The repercussions of enhanced efficiency—streamlined operations, reduced costs, and improved productivity—come with a challenging counterpart: the potential personnel displacement. It's the paradox of progress: the innovations that propel us forward can also leave some of us behind.

While the appeal of AI efficiency and the potential for leaner operations can be tantalizing, the implications for your workforce cannot, and should not, be ignored. Increased efficiency often translates into fewer human hours needed, directly threatening employment. While economically advantageous in the short term, the allure of a smaller, highly efficient workforce may cast a long and ominous shadow over your team, sowing seeds of anxiety and mistrust. "Who's next?" might become the unspoken question haunting every office and shop floor.

The initial impulse might be to lean into these newfound efficiencies, to pare down the workforce in pursuit of a leaner, meaner operation. And yes, from a short-term perspective, such a path might lead to an upswing in profits. But we must never forget that business, like life, is a marathon, not a sprint. The long-term view often reveals that what seems beneficial in the immediate can prove detrimental down the line. In the context of workforce reduction, the immediate savings could be offset by future costs: the challenges of re-hiring during a growth phase, the loss of institutional knowledge, and the impact on the morale of the remaining employees.

Here, we are faced with a delicate balancing act. On the one hand, we have the drive towards automation and efficiency; on the other, we have the human element—the need to protect our employees, their livelihoods, and their morale. How does one reconcile the two?

Firstly, we must accept that workforce reorganization is not simply a cut-and-dry exercise in numbers—it's about people. Reducing headcount may seem straightforward, but it has implications beyond the balance sheet. The erosion of employee morale and trust, the tarnishing of your organization's reputation, and the challenge of attracting talented personnel during growth phases are considerable risks that come with a hasty workforce reduction.

Every employee is an ecosystem, a complex web of skills, experiences, relationships, and knowledge. When one such ecosystem is lost, it's not just a personnel number taken off the books; it's a slice of the organization's collective memory, a conduit of its culture, and a carrier of its values. Moreover, the fear of layoffs can cast a long shadow over an organization, breeding uncertainty and distrust, hampering productivity, and eroding company culture.

So, how can organizations navigate this delicate balancing act between efficiency and empathy? The answer lies in foresight, adaptability, and a deep commitment to the most valuable asset of any organization: its people.

In the next few pages, we will explore specific strategies that strike the elusive balance between AI-led efficiencies and workforce wellbeing. From upskilling initiatives and role redefinition to job rotation and creative collaboration, these strategies aim to foster a culture of learning and evolution where AI and humans coexist, complement, and even catalyze each other's potential. We will explore how companies can foster a continuous learning and adaptability culture, ensuring employees survive the AI revolution and thrive in it.

Ultimately, the true measure of an organization's success in the era of AI won't be just about bottom lines and productivity metrics. It will be about whether they've created a win-win scenario: a future where the company and its people grow with it. A future where AI serves as a springboard for human potential, not a stumbling block. A future where progress, in its truest sense, is progress for all.

> **We will have a very small pool of senior people if entry-level jobs are replaced with AI.**

The "Man – Machine – Map" model discussed earlier provides a critical lens through which we can view the shifting landscape of the workforce. In this model, Quadrant 4, which we may call the "Automation Zone," represents roles that AI is increasingly taking over. These are typically roles involving repetitive tasks, routine problem-solving, or data crunching—activities that machines excel at. It's also, unfortunately, where job displacement is most likely to occur as AI capabilities advance.

However, viewing this as merely a displacement problem would be missing the bigger picture. This transformation also presents a massive opportunity to rethink our approach to workforce development and job allocation. Suppose we can proactively identify the roles falling into this Quadrant. In that case, we can begin strategizing how to transition these employees into roles where human skills are highly valuable.

As pointed out, the first direction of this transition is towards Quadrant 1: the "Fusion Frontier." Here, human and machine skills complement each other to achieve new heights of performance. Roles in this Quadrant would involve employees using AI to amplify their skills and increase their productivity. For example, a data analyst equipped with AI-driven data processing tools can analyze massive datasets more accurately and in a fraction of the time, leading to more informed business decisions.

Reskilling employees for these roles would involve training them in using AI tools and strategic thinking, creativity, and complex problem-solving that can fully leverage the power of AI. These are skills that machines cannot replicate and are, therefore, highly valuable in the age of AI.

The second transition direction is towards Quadrant 2: the "Human Heartlands." Here, the focus is on roles that require uniquely human skills like empathy, creativity, leadership, emotional intelligence, and understanding and navigating social contexts. These are roles that, by their nature, are resilient to AI disruption. Jobs in healthcare, education, social work, leadership, and creative industries fall into this category.

The pathway to this Quadrant involves training employees in these human-centric skills, fostering their emotional intelligence, nurturing their leadership capabilities, and honing their creative abilities. The beauty of this Quadrant is that it offers roles that are not just secure in the face of AI disruption but are also fulfilling on a human level.

Conducting a workforce audit from the Human-Machine skills axes model perspective can provide a strategic roadmap for future-proofing your workforce. It can help illuminate the path towards a future where your employees are not victims of AI advancement but beneficiaries, harnessing its power to improve their skills, boost their productivity, and find greater fulfillment in their work.

Recalibrating Complementary Activities

In the earlier chapter on technological evolution, we discussed how technologies can be either "Transformative or Frontier" in their nature. This provides an interesting perspective and potential solutions in the context of "Workforce transformation" with advancing AI technologies. This dual lens of "Transformative" and "Frontier" technologies serves as a beacon to guide businesses in the era of AI. By understanding these facets, enterprises can leverage AI to its fullest while making strategic decisions that ensure growth and human resource well-being.

Let's first delve into the concept of "Transformative" AI technologies. These technologies do not invent new tasks; rather, they automate existing ones, making them more efficient, accurate, and scalable. They are akin to a digital assembly line, tirelessly performing tasks, sifting through mountains of data, identifying patterns, and providing insights. In the business context, it's imperative to understand that AI, while extraordinarily powerful, is fundamentally a tool. It can't dream. It can't imagine a future that doesn't exist. And it can't inherently break out of its programming to create something entirely new. However, it can efficiently and effectively handle repetitive tasks, process vast amounts of data, and identify patterns humans might overlook. AI's true power lies in its ability to transform the traditional tasks within your business to free up your most valuable resource - the human mind.

For instance, in a customer service department, AI chatbots can handle many basic queries. At the same time, sentiment analysis algorithms can process customer feedback to identify general trends. In manufacturing, predictive maintenance algorithms can predict equipment failures, reducing downtime and maintenance costs. All of these are tasks humans have traditionally done, but AI can do them faster, better, and on a larger scale.

Imagine, for a moment, a modern, bustling insurance firm. An army of employees diligently answers customer queries, manages claims, underwrites policies, and analyzes risk data. The workload is immense, and the tasks are often monotonous. Now, let's inject AI into this picture. A sophisticated AI chatbot handles routine customer queries. Automated claims processing systems manage straightforward claims. Predictive algorithms perform risk analysis, and AI-powered systems assist underwriters. The result? A transformed operation that executes routine tasks with increased speed, efficiency, and accuracy. Applying Transformative AI technologies effectively offloads routine, repetitive tasks from the human workforce. The benefits of this automation can be naively and quickly equated to cost savings and economic efficiency.

However, the true magic of this transformation isn't the newfound productivity or cost savings, though these are certainly valuable. The real game-changer is that you've freed your human workforce from the daily grind of routine tasks. You've created the time and space for them to dream, innovate, and explore the frontiers of your business.

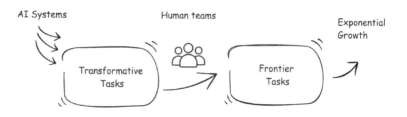

This brings us to the second part of our lens - the "Frontier" tasks. These are new, previously unexplored areas that AI opens up for businesses. Unlike Transformative tasks that are largely about efficiency and scale, Frontier tasks are about novelty and creating new products, services, or processes that couldn't have been envisioned.

Consider our previous example of the insurance firm. With AI handling much of the customer service load, human resources are now available to explore Frontier tasks. These tasks require human creativity, strategic thinking, and understanding social contexts - qualities AI doesn't possess. Here, our insurance firm's employees can devote their time to developing unique, cutting-edge insurance products tailored to niche markets. They could investigate and capitalize on emerging trends like micro-insurance and on-demand insurance or even explore new domains. Their focus could shift towards enhancing the customer experience through streamlined service and deep, personalized engagement.

The firm could also explore creating new processes that enhance the customer experience. For example, AI could be integrated into the firm's services, allowing customers to customize their insurance plans or easily file claims through an AI-assisted platform. The firm could even pioneer a new way of offering insurance services, setting a new industry standard.

The Transformative-Frontier lens offers a strategic framework for businesses in the era of AI. By applying AI to Transformative tasks, businesses can significantly increase operational efficiency and free up human resources. These resources can then be redirected toward Frontier tasks, which offer innovation, growth, and market leadership opportunities. The result is a business that is not only more efficient but also more creative and forward-looking. This business is fully prepared to thrive in the age of AI.

By adopting AI to take care of the transformative aspects of the business, we can redirect human ingenuity towards frontier work, unlocking possibilities that we've never imagined. It represents a significant evolution of the very concept of work, moving away from the task-oriented paradigm to a purpose and innovation-driven model.

Furthermore, imagine the positive impact on employees' morale and satisfaction. Liberated from the mundanity of routine tasks, they can apply their creativity, strategic thinking, and human insight to forge new paths. By reskilling employees to focus on Frontier tasks, businesses retain their workforce and provide them with opportunities for growth and fulfillment. This kind of work is not just economically valuable; it's personally fulfilling and aligns with a higher purpose - two crucial factors in employee engagement and retention. By approaching AI transformation through this lens, businesses can ensure their success in the era of AI while also caring for their most important asset - their people.

Wrapping up, it's more than clear that it is an exciting time to be in business in this nascent beginning of AI

7

CRAFTING THE DISRUPTION: INVESTORS AND ENTREPRENEURS

Investors hold a unique and powerful position in this thrilling era of AI. They serve as the springboard for innovation, as the conduits that propel startups and enterprises from ideas to reality. But being the custodians of capital, they also bear the responsibility of chiseling the course of AI's evolution. Their task goes beyond mere transactions and into sculpting the future. Entrepreneurs from the other side of the same coin.

Every aspect that applies to the investing world applies equally to the world of founders and innovators.

But with great power comes great trepidation. A paradox stares investors in the face – on the one hand, they crave the certainty of well-established, low-risk businesses, and on the other, they cannot ignore the compelling potential that AI promises. It's like standing on the edge of a precipice, gazing at a landscape of limitless possibilities, yet grappling with the fear of the unknown. This conundrum births a fear of missing out (FOMO) that could steer investors toward a tumultuous sea of uncertainty.

Being an investor in the AI revolution is akin to standing in a maze with a thousand doors. Behind each door lie worlds with potential returns and unprecedented growth cloaked with unknown technological hurdles, market acceptability challenges, and unchartered regulatory landscapes. In this maze, an analyst or partners in a venture capital (VC) firm might find themselves lost, their judgment clouded by the fog of hype, or misled by a trail of loosely connected pieces of information scattered across the internet.

This is where understanding the nuances of AI becomes an investor's beacon. An intimate understanding of the principles of AI, its capabilities, and its limitations becomes a compass to navigate this labyrinth. Investors need to equip themselves with a bird's eye view of the big picture while cultivating the insight to delve into a molecular view when necessary.

Striking a balance between these perspectives is an art that needs to be mastered, akin to simultaneously seeing a grand chessboard and being able to strategize each move.

Earlier, we traversed the vast terrain of AI's nature, evolution, and disruptive power on the nature of work. This chapter will extend this exploration, focusing on the nuances an investor needs to understand. We will dive into the depths of risk assessment, returns, and investment timelines, unraveling layers of knowledge to bring clarity amid the wild west, the current stage of AI investment.

The current investment landscape is awash with a 'Spray and Pray' approach, indicating the chaos when the fear of missing out overtakes rational decision-making. This chapter seeks to mitigate that stress by illuminating the path for investors and providing them with the tools to decipher the cryptic world of AI and use an "Aim and Play strategy."

As investors, it's crucial to remember that you are not mere spectators in the theatre of AI. You are architects shaping the foundations of the AI revolution. Your decisions will not only yield economic returns but will have a lasting impact on humanity's capabilities. Your role is both pivotal and profound.

Embark on this journey of understanding AI through an investor's lens. We'll navigate the convoluted corridors of risk and reward, understand the interplay of technology and regulations, and lay the groundwork for making informed and impactful decisions. Your guidebook begins here and promises to take you through the labyrinth of AI investment onto a path of clarity and insight.

Remember, you're not just investing in technology; you're investing in the future of humanity.

Ephemeral Supernova or an Eternal Galaxy

Like sailors once used the stars to navigate uncharted waters, investors use previous technological epochs as their guiding constellations. The transition from the age of the Industrial Revolution to the Information Age and then to the Digital Age has, over time, painted a familiar pattern on the canvas of progress. Today, we find ourselves on the cusp of the AI epoch, standing at the edge of a precipice, once again ready to leap into the unknown.

Each previous technological cycle, like the rising and setting of celestial bodies, has provided us with invaluable lessons. It's like a journey through a time-lapsed galaxy, where each previous revolution – from steam engines to the internet – is a star illuminating the pathway for investors. However, The AI epoch is akin to discovering a new constellation that demands we reinterpret our charts and learn a new celestial language.

Draw parallels between the current AI excitement and the recent hype around blockchain and cryptocurrencies. Indeed, the fervor shares similarities. There's the same dizzying jargon, the same evangelists claiming a world-changing revolution, and the same fear of missing out that fuels a frenzy of investment. The recent hype cycle of blockchain and cryptocurrencies felt like the discovery of a 'gold rush,' a fever that drove many towards quick fortunes. It was a narrative filled with promises of decentralized power and unprecedented transparency. However, this narrative often eclipsed its complexity, volatility, and limited scope of impact.

It was as if we had discovered an exciting new star, only to realize it was a supernova – dazzlingly bright but inherently unstable. But here's where the analogy ends. The blockchain boom was largely speculative, built on an expectation of the value that decentralized digital currencies could bring in a potential future. Conversely, AI is not merely an expectation or a promise; it's a reality unfolding right before our eyes. While cryptocurrencies offer a different way to transact value, AI is here to redefine value creation itself.

The AI revolution is no ephemeral supernova. It represents a sea change, more akin to the birth of a new galaxy. It is a technology that will alter how we interact with machines and each other. If blockchain were a new currency, AI would be a new language.

Like the emergence of spoken language, which propelled our ancestors from simple tool users to sophisticated societies, AI is redefining the interface between humans and technology. AI's potential transcends hype because it's not just about creating a new form of currency or a new way of doing business. It's about crafting a new form of understanding, a new way of interacting, and, ultimately, a new way of being.

Suppose the internet age was like harnessing electricity, powering homes and industries. In that case, the AI revolution is akin to harnessing fire. It is elemental, transformative, and can be both beneficial and destructive. As investors, your role is not merely that of a bystander. It's akin to being a firefighter. You control this powerful force, ensuring it brings light and warmth rather than destruction.

This chapter will continue to demystify the nebulous terrain of AI investment, showing you how to navigate this new galaxy. We will illuminate the pathways, highlight the constellations, and help you chart a course through the AI revolution. Like explorers who sailed uncharted seas guided by the stars, investors need to be empowered with the knowledge and understanding to navigate the AI epoch. You are part of the crew and an essential co-captain of this journey.

As we sail together, remember our goal is not merely to conquer new lands but to shape them for the benefit of all humanity.

In our preceding discussions on understanding AI, we've embarked on a riveting journey through time. We've traced AI's evolution, grappled with its present challenges, and peered into the vast expanse of its future potential. With this knowledge, we are poised to demystify and delineate the intricate AI ecosystem. It's akin to plotting constellations in the vastness of the cosmos; seemingly random, but with careful observation and understanding, a pattern emerges.

The complete AI landscape, keeping aside the hardware aspect, can be categorized into four distinct clusters:

1. **Foundation Models and Core Algorithms**

2. **Development of Models on New Domains of Data**

3. **Applications of Models in Specific Industries and Use Cases**

4. **Integration of These Technologies into Existing Products and Features.**

Why is this classification crucial for investors? Just like a traveler setting sail on uncharted waters needs a compass, an investor venturing into the AI landscape needs a map to navigate its complexities. This categorization serves as that vital map, highlighting potential investment territories and providing an understanding of the underlying terrain.

The first category—Foundation Models and Core Algorithms—is the bedrock of AI development. They are the 'basic physics' of this AI universe, the fundamental principles on which every AI' planet' orbits. This terrain is the bedrock of AI. It's the fertile ground where the seeds of AI are sown and cultivated. It's the realm of mathematicians, computer scientists, and algorithm designers constantly pushing the boundaries of what's possible. By understanding these models and algorithms, investors gain insight into the genetic makeup of AI, its strengths and limitations, and the potential for groundbreaking discoveries.

The second category—Development of Models on New Domains of Data—delivers the key to unlocking untapped territories. It presents the frontier where AI extends its reach, adapting and evolving to suit new terrains. This terrain is the crossroads where AI meets real-world data. It intersects theory and practice, where foundational models are adapted and honed to navigate uncharted data territories. For an investor, understanding this segment is akin to unearthing a treasure trove of potential, revealing where next-generation AI' species' might evolve.

Next, we have Applications of Models in Specific Industries and Use Cases. This is where the rubber meets the road, where AI leaps from theory into practice. It is the marketplace where AI solutions compete, transform industries, and reshape our lives. This is the terrain where AI comes alive.

It's the bustling marketplace of AI, where models find meaningful applications, solve real-world problems, and add tangible value to industries and users alike. For investors, this represents the 'trade routes' of the AI realm, connecting solutions to industries, catalyzing growth, and ultimately yielding returns.

Finally, the Integration of These Technologies into Existing Products and Features category represents AI's seamless fusion into the fabric of existing systems. It's the AI equivalent of the Industrial Revolution's assembly line—pervasive, transformative, and integral to the operations of tomorrow's world. This final terrain is the bridge that connects AI's potential to its practical utility. It's the realm of software engineers and product designers, weaving the power of AI into the fabric of products and features we use daily—the bridge made up of APIs connecting myriad end-user applications to core AI system providers.

Some companies are limited to a single terrain, and some expand over multiple terrains. Each terrain needs specialists in research, development, marketing, and support requirements. This four-pillar classification helps investors understand the breadth and depth of the AI landscape. It'sIt'se a map, revealing the interconnections and dependencies between these terrains. It provides investors with a holistic view of AI'sAI'selopment world, enabling them to identify synergies, evaluate risks, appreciate potential, and gauge timelines for maturity.

This will provide a clear lens through which you view the AI landscape. As we progress, we will sharpen this lens, bringing the world of AI into increasingly sharper focus.

AI TECHNICAL LANDSCAPE FOR INVESTORS

FOUNDATIONS

1. Efficiency & Specificity

2. Online learning

3. Explainability

4. Transparency & Bias

DATA DOMAINS

1. Embeddings & Vectors

2. Multimedia

3. Games, Architecture, ...

4. Fashion, Synthetic biology, ...

ENTERPRISE APPLICATION

1. Tailoring for Industries

2. Consolidation of AI Services

3. Intelligent Data Management

4. Development tools

CONNECTORS - API

1. Healthcare

2. Education

3. Publishing

4. Travel & toursim, ...

5. Legal & GovernancE, ...

Foundations of AI - The Core and The Quandary

Our exploration of the AI universe begins with its celestial core - the foundational models and algorithms. Much like the fiery center of a galaxy, this domain is the heart of AI development, pulsating with raw power and immense possibilities.

This is where the laws of AI physics are written, shaping the dynamics of all downstream domains. Yet, it is also where we encounter our first significant challenges and opportunities, a testament to the fundamental principle that every new frontier presents obstacles and untapped potential.

One of the most notable challenges is the daunting twofold task of 'Efficiency and Specificity.' Let's delve deeper into what these terms mean in the context of AI. Imagine an AI model as a colossal ship, its capabilities equivalent to cargo. To journey through the sea of data and deliver its 'cargo,' this AI ship needs a staggering amount of fuel or energy and computing power in the world of AI.

Consider the widely used AI model, ChatGPT. It demands an astronomical 1000 MWh of energy for training and about 260 MWh daily to operate. This is akin to the power consumed by small towns just to keep a single AI model operational! In this context, efficiency reduces this prodigious energy consumption to make AI models sustainable and environmentally friendly.

And then there's 'Specificity'. Just as every sea voyager doesn't need to explore every inch of the ocean, does an AI model need to be trained with all the data in the universe? The current trend is akin to trying to feed the entire Library of Alexandria to an AI model when all it might need is a single scroll.

So, one exciting avenue is to steer AI development toward 'specificity.' Imagine AI systems akin to human scholars. They start with a good understanding of language (or the foundational knowledge) and then build specific expertise in different domains. An AI legal expert, for instance, does not need to grasp the intricacies of Victorian architecture or delve into the nuances of Bohemian art. By focusing on specific fields, we could control the knowledge the AI model needs to master, leading to more efficient models that consume fewer resources.

In this light, 'specificity' and 'efficiency' become the twin pillars to address the challenges of energy consumption and excessive data training. They could pave the way for a new generation of AI models that are as knowledgeable as they need to be, not as knowledgeable as they could be. This shift could usher in a wave of market competitiveness as more specialized and efficient AI models begin to dominate the landscape.

This pivot towards efficiency and specificity is one of the many facets of the foundation and core algorithms domain we'll explore. As we voyage further into the AI universe, we'll navigate through more of these challenges and opportunities. Like explorers charting new lands, we'll come to understand that every challenge presents an opportunity to innovate, and every innovation carries the potential to reshape the AI landscape.

Power of Online Learning - The Path to AI Enlightenment

As we continue our expedition into AI's heart, we face our next monumental challenge and opportunity—online learning. This feature, or rather, the lack thereof, in AI models today is akin to an astronomer's telescope fixed on the past, unable to perceive the unfolding cosmos. Even the most advanced AI models today are trained on past data, with a strict cut-off date, making them oblivious to the steady stream of new information generated every second.

Just imagine the predicament of an AI system trained on data up to 2022 but has no cognizance of events or knowledge beyond that. For instance, the AI system wouldn't know about the latest advancements in quantum computing, political changes, or the release of the newest iPhone model. It's like a time traveler from the past trying to navigate the present world: able but incomplete, knowledgeable but outdated.

Currently, this issue is sidestepped by integrating the AI models with the internet through search engines, providing a window to the most recent data. However, this is a workaround, not a solution, because the new data is not assimilated into the AI's core understanding. This limitation results in AI models that, over time, might resemble an archaic artifact more than a cutting-edge technology.

The solution to this problem lies in empowering AI models with online learning—the ability to continuously learn and update their knowledge base with newly generated information. Consider this equipping our AI astronomer's telescope with the ability to time-travel, to perceive not just the past but also the unfolding present and the hint of the future.

Think of the edge an AI model with online learning could provide, especially at the enterprise level, where internal data is continuously generated. It's like having a personal scholar who never stops learning, keeping up with your specific field, assimilating new information, and providing relevant insights. Over time, such an AI

model could become an irreplaceable asset, an intellectual companion that evolves and grows with your enterprise.

However, this promising prospect of online learning comes with its own set of challenges. One significant concern is to prevent AI models from getting lost in the recursive loop of being trained over AI-generated content. It's a complex issue, akin to ensuring our AI astronomer doesn't get lost in a black hole of its own making.

The right investments can guide us towards a future where AI is not just knowledgeable but also an ever-learning intellectual companion.

Enigma of AI: Explainability, Transparency, and Bias

Let's venture into another remarkable domain with significant bearings on AI's evolution and acceptability—the mystifying realm of "Explainability and Transparency" and the intricate labyrinth of "Bias." Let's navigate through this intriguing world where science intersects ethics, transparency grapples with opacity, and technology rendezvouses with sociology.

In the early days, algorithms were hand-crafted. Each step and decision within the system could be meticulously tracked and analyzed. For instance, scientists would write specific rules and criteria based on distinctive facial features when programming a system to detect human faces. Over time, the system was fine-tuned with data to increase accuracy. It was akin to a well-planned city where every lane and building had a designated role, and the city planner knew exactly why and how each element functioned.

However, with the advent of Deep Neural Networks (DNNs), the city planning approach was replaced by a rainforest's chaotic growth. In DNNs, the features and rules are not handcrafted but emerge through learning from vast amounts of data. While incredibly powerful, these systems are opaque, giving rise to the term 'black box'. In the urban metaphor, it's as if we have a sprawling, ever-growing city where buildings sprout spontaneously, roads wind in unpredictable directions, and yet somehow, it all works, most of the time.

Much like an expert chef who effortlessly blends ingredients to create a gastronomical marvel but struggles to explain the nuanced process, AI models, particularly deep neural networks, have become adept at delivering astounding results without the ability to elucidate the logic behind their outcomes. While delivering high precision, the complex interconnected web of computations remains a black box – an enigma.

This obscurity in AI models presents a challenge when they fail or produce unexpected results, for it is often difficult to understand the cause. There is no backtrace of the thought process, no notes in the margin, and no explanation. Imagine our AI city producing cars that sometimes, inexplicably, turn left instead of right, and we have no idea why.

These opaque models, although impressive in their execution, become a double-edged sword when results need to be interpreted or repeated. Imagine an AI system assisting doctors with medical diagnoses – it's not enough to know the diagnosis; understanding the reasoning behind the diagnosis is equally, if not more, important. What factors did the AI consider? What patterns did it observe? In such scenarios, the "why" becomes as critical as the "what." This transparency of AI models, or the lack thereof, thus poses a significant challenge to their large-scale implementation, especially in sensitive domains.

Furthermore, staring into the AI mirror reflects our intellectual capacity and biases that lurk in our subconscious. Bias in AI, an inadvertent inheritance from the data it's trained on, is another key challenge akin to an invisible enemy.

This problem is compounded when considering the societal consequences of "Bias in AI." AI systems learn from the data they are trained on. If this data carries human bias, the AI system can and often does, inherit these biases. Let's return to our city metaphor. The builders – the data – may all come from one demographic. They have built houses, restaurants, and schools reflecting their taste and lifestyle. Consequently, the city reflects this demographic, favoring its needs and preferences over others.

Biases can enter the training data in two primary ways: through the data created by humans and through those who curate this data before training. Unconscious biases ingrained in society can find their way into AI systems, creating a troubling reflection of societal prejudice. An AI system trained on biased data could perpetuate and amplify existing inequalities, whether racial, gender-based, or socioeconomic. Much like a language learner who inadvertently picks up the slang and accent of their tutor, AI systems can pick up and amplify the biases present in their training datasets, which are, after all, the creations of human beings replete with their conscious and unconscious biases. Curators who prepare these datasets can further compound this bias.

Bias in AI can have profound and far-reaching consequences. Consider an AI system used for hiring – a biased model could perpetuate discrimination and hamper diversity, or even worse, create an unfair playing field in sensitive domains like law enforcement or credit approvals. In our metaphorical city, if most buildings were designed with tall stairs and no wheelchair ramps because the builders didn't consider wheelchair accessibility, it would be unwelcoming and inaccessible for people using wheelchairs.

Attempts have been made to circumvent these problems by filtering some terms in the output data or "sanitizing" the input data. But with AI models capable of handling and generating billions of question-answer combinations, filtering is a mere band-aid on a deeper issue. It's as if, realizing the city is inaccessible to wheelchair users, the authorities are trying to fix the problem by banning cars. This measure doesn't address the root of the issue. This task is not just a technical hurdle but also an ethical one. It involves questioning how we, as a society, want AI to mirror us. It's about remodeling our city to be inclusive, accessible, and fair.

Solving these fundamental challenges—achieving AI explainaility, ensuring transparency, and eliminating bias—will be a breakthrough akin to finding the proverbial needle in the haystack. But the opportunity it presents is enormous. Any organization that navigates these waters will stand at the vanguard of the AI revolution, wielding the power to reshape entire industries and rewrite business rules.

Investors, then, must be acutely aware of these aspects. They should look for companies that demonstrate mastery of AI's capabilities and show a deep commitment to cracking the code of explainability, championing transparency, and actively combating bias. In the grand scheme of AI evolution, these attributes are the most lucrative markers of a promising investment.

Dispelling the Illusions: AI 'Hallucinations' and the Quest for Reliable Predictions

Just as we sometimes see shapes in clouds that aren't truly there, AI systems can hallucinate, generating or predicting information that isn't based on their input data. This is not a product of an AI's imaginative capabilities but is a symptom of the system's over-reliance on its training data and the patterns it has learned. The AI systems, over-reliant on their learned patterns, begin to make presumptions even in the absence of these patterns in new data. It's akin to a seasoned detective seeing clues in every shadow, a consequence of being steeped in the intricacies of their profession for too long.

Hallucination in AI often surfaces when we venture into unchartered waters, asking questions or presenting scenarios that stray from the training data's path. This occurrence is often prompted when users compel the system to venture into uncharted territory – asking questions beyond the AI's training data realms. In such cases, the AI generates seemingly plausible but ungrounded responses from low-probability associations. Picture asking a home assistant, designed for personal use and trained on consumer data, to provide insights into quantum physics. The assistant might conjure up some related terms, formulating an answer that seems plausible but isn't based on actual knowledge or understanding.

This problem can spiral into complex repercussions, mainly when accuracy is non-negotiable. Picture a medical AI diagnosing a patient based on non-existent symptoms because it hallucinated a connection without any, or a financial AI predicting a market crash based on phantom correlations. The results could be catastrophic.

Yet, as we have seen throughout our exploration of AI, each challenge presents its own set of opportunities. This issue of hallucination is no different. Advancements that curb the tendency of AI to hallucinate can pave the way for more reliable and robust models, distinguishing more accurately between the patterns they've learned and the data they're currently processing.

This brings us to another crucial attribute that could significantly enhance the effectiveness of AI systems: the capability to recognize the 'unknown,' or as we call it, "Ignorance Awareness." A valuable trait in human intellect, knowing what one does not know, can be a game-changer in AI. This profound concept transcends the binary world of known and unknown into a space where the system understands and admits its limitations. Much like a humble human expert who isn't afraid to say, "I don't know," an AI system that can discern when it's straying beyond its training and express that to the user would be a giant leap toward more reliable and trusted AI interactions. Much like a seasoned scholar, an AI system that can admit, "I don't know, but I can try," would be significantly more trustworthy and user-friendly. This transparency allows users to evaluate the AI's responses critically. It protects them from being misled by confident yet unfounded outputs.

Investors should keep companies pioneering solutions to these complex issues—hallucinations and ignorance awareness—on their radar. The ability to tackle these challenges could unlock an entirely new level of dependability in AI systems and open up immense market potential. Companies prioritizing these aspects are also likely to be more resilient in the face of the evolving AI landscape and better equipped to navigate the challenges that come with it.

This is not just about developing smarter technologies but about the birth of systems that can engage with us more authentically and responsibly. Remember that while we are looking for AI systems that can provide answers, breakthroughs will also come when they can admit what they don't know. We must look beyond just performance statistics and heed these subtle yet profoundly impactful facets of AI development. They could be the critical differentiators between an ordinary investment and a profoundly transformative one.

Uncharted Horizons - New Domains of Data

Imagine standing on the brink of a vast, undiscovered land, teeming with possibilities and mystery. As an explorer, your pulse quickens at the prospects. Each hill and valley is a new form of data waiting to be discovered and understood. The world of AI is no different, presenting a similar, exhilarating frontier— the exploration of new data domains.

The leaps we've made with AI's first terrain, the 'Foundation,' have equipped us with a versatile, powerful tool - Deep Neural Networks (DNNs). DNNs are like the all-terrain vehicles of the AI world, capable of transcending data domains. They can adapt and tackle any domain of data we set them upon - from text to images, from audio to video, even to the intricacies of software code, you name it.

This flexibility rests on the innovation of diverse architectures built on these deep neural networks, making handling a wide range of data modalities possible. However, driving this vehicle requires a map, a guide to navigate new data's vast and complex terrain. This is where the second terrain comes into play - building that map, devising new architectures, and creating structures to translate these different data modalities into formats DNNs can comprehend.

This terrain is the playground of computer scientists, mathematical geniuses, and algorithmic wizards. Like pioneering gold prospectors, they push the frontiers, converting uncharted territories into rich, mineable landscapes. These pioneers work with a tool known as "embeddings," turning data of any form into "vectors," a universal language that DNNs understand. It's like the alchemical process of transmutation, taking base materials and turning them into gold, unlocking a myriad of applications in the process.

Much like early explorers mapping uncharted lands, these innovators translate the complex landscape of new data into digestible, useful formats.

So, why is this relevant to investors? It's simple. Each new data domain successfully mapped and made understandable to DNNs unlocks an enormous range of potential applications. Just like how the discovery of new lands led to unprecedented opportunities for trade and economic growth in human history, opening up new data domains to AI will unlock untapped market potentials and innovative products.

Venturing into these new data domains signifies entering a realm with opportunities, much like an investor discovering an untapped market. As each new data domain unfolds, it brings along an array of applications, each with the potential for substantial returns. It's akin to opening a treasure chest and finding it full of countless sparkling gems. Each gem signifies a novel application, and each application promises remarkable returns.

Imagine the possibilities when DNNs can understand "architectural floor plans" on an unprecedented scale or when they can "generate new car designs" based on text descriptions. Every new domain conquered creates a new market opportunity. For investors, recognizing and investing in the ventures pushing the boundaries of these uncharted data domains could yield astronomical returns.

Investors should view these data domain explorations with an eye on the future. These are not just algorithmic innovations; they represent new market segments, product lines, and revolutionary services. They signify the capability to unlock value from unprocessed data, much like a refiner extracting precious metals from raw ore.

Each new data domain conquered could translate into an array of AI-driven solutions, creating a ripple effect of value generation. Thus, as investors, keeping a close watch on these new domains and the companies spearheading these explorations could open doors to exciting, high-growth opportunities. As investors, keeping a keen eye on the horizon is critical, looking out for those venturing into these new frontiers. For in the world of AI, the adage rings true - fortune does indeed favor the bold.

Media Frontiers - Exploring 3D, Video, and Audio Domains

Dare to picture a world where you can describe a sculpture to an AI, and it creates a 3D model within minutes. Or consider a scenario where you outline a brief scene to a program, and it generates a high-quality video clip as per your vision. The frontiers of AI, though nascent, are pushing the boundaries of possibilities, and the unchartered territories of 3D, Video, and Audio data domains promise a leap in our imaginative potential and creative capability.

Take the 3D domain as an instance. Today, creating a 3D model involves a highly skilled human hand utilizing sophisticated software to meticulously design, shape, and render the model. It's time-consuming and demands an expert touch. But with advances in AI, we stand on the brink of a revolution. Imagine an AI system where you describe a character or object, and the system generates a fully formed 3D model that can be instantly integrated into video games, movies, or virtual reality environments. This could revolutionize the gaming, architecture, medicine, and education industries, drastically reducing production time and costs and democratizing access to 3D design.

In the video domain, AI has the potential to redefine the way we create and consume content. Producing high-quality video or animation requires considerable resources, time, and skill. But what if we could generate video content just by describing a scene? AI could automate video production, making it more accessible and affordable for movie studios, small businesses, educators, and individuals. The implications for entertainment, advertising, and education are vast and untapped.

Now consider the audio domain. Today, synthesizing natural-sounding human speech or music is a complex task. However, AI advancements could enable us to generate custom speech or music merely by providing a textual or musical description. Imagine producing a symphony without touching an instrument or creating a podcast with natural-sounding voices without hiring voice-over artists. The potential here extends to the music, broadcasting, and digital marketing industries.

These nascent 3D, Video, and Audio domains are still in their early stages. Translating these complex data forms into a language that AI understands is no mean feat. But the potential is enormous. These innovations are not just about time or cost savings; they represent a paradigm shift in our creative process. They enable us to generate rich, immersive, and interactive content at a scale and speed that was previously unimaginable.

As investors, these burgeoning domains represent significant opportunities for astronomical growth. The ventures pioneering in these fields might be in their infancy, but their maturity promises significant returns. In the AI investment landscape, these domains are akin to promising gold mines—still requiring a lot of digging and sifting but potentially hiding vast fortunes.

As we voyage deeper into the AI revolution, let us take a moment to delve into our imaginative playgrounds - video games and architectural design. These creative fields are set to be reshaped by

AI in remarkable ways that are nearly impossible to anticipate today fully. However, let's dare to dream and visualize a future where AI assists us in crafting our fantasies and realities.

Imagine, for a moment, an AI Game Master - an advanced AI tool designed to assist in creating intricate, immersive, and interactive gaming experiences. You begin by simply explaining the world you want to create. You want a fantasy realm lush with unique flora and fauna, filled with mystic creatures and ancient ruins. In moments, your AI companion generates an intricate 3D map of your world, accurately bringing your description to life.

Now, picture designing characters. You describe an enigmatic elf queen, vivid in your imagination. The AI translates your words into a lifelike 3D model, down to the last detail of her iridescent armor. From the grandeur of epic landscapes to the minutiae of character designs, AI can assist, enhance, and accelerate game development, freeing human creators to focus on the heart of their games: storytelling.

The AI Game Master doesn't stop there. It can generate original music scores based on the mood of different game scenes, devise challenging puzzles for players to solve, and even help design the game's plot. It adapts the storyline in real-time, reacting to players' decisions and actions, ensuring an immersive, personalized experience.

Now, let's turn our focus from virtual realities to tangible ones - architectural design. Picture an AI Architect tool with an extensive architectural knowledge database, design principles, and regulations.

You feed it your brief: a sustainable housing complex designed to encourage community interaction and optimized for energy efficiency. The AI almost instantly generates drafts with 3D models, floor plans, and detailed project specifications. It takes care of the mundane drafting tasks, allowing the human architect to invest their energy in refining the design, optimizing space usage, and enhancing aesthetic appeal.

The AI Architect doesn't just adhere to the brief; it suggests improvements based on environmental data, building orientation, local climate, and more. It can even predict and visualize the effects of natural elements on the structure over time, allowing for proactive design modifications. All this while the human architect steers the creative process, making the key decisions.

The transformative impact of AI on these creative industries is staggering. As AI technologies mature and become more integrated with these sectors, they could democratize the creation process, accelerate development timelines, and inspire new levels of creativity. For investors, backing these technological frontiers could yield substantial financial returns and the satisfaction of participating in a paradigm-shifting revolution.

The future of AI lies not just in its computational power but in its ability to inspire human imagination and creativity. And that, indeed, is a bet worth taking.

Enter the vibrant world of fashion design, where AI has the potential to take center stage as the ultimate designer's assistant. Picture this: an AI-driven design software that can understand and predict fashion trends based on social media analysis, runway showcases, and historical data. This AI Fashionista can quickly generate hundreds of design sketches, blending new trends with classic styles.

Switching gears, let's venture into the awe-inspiring world of synthetic biology. Here, AI could play a critical role in designing new biological systems. Picture a future where we can "program" biological organisms like we code software today. AI-driven design tools could create custom DNA sequences for new organisms optimized for specific tasks.

AI's capability to venture into new domains and reimagine traditional industries is remarkable. It's bringing about a paradigm shift across sectors, and this change is only in its infancy. Architecture, Gaming, fashion and synthetic biology examples we explored are mere drops in an ocean of possibilities. With AI, the uncharted territories are vast, filled with untapped potential waiting to be harnessed.

Investors looking to pioneer change, influence the future, and garner substantial returns should look for AI's next big foray. These burgeoning areas represent a massive opportunity for progress and a sound investment. Stay curious, and most importantly, stay ready to seize the opportunities AI continues to unveil.

Conquering New Territories – Enterprise Applications

As we navigate further into the vast terrain of AI, we encounter a pivotal domain where the rubber meets the road: the application of AI across various industries and specific use cases.

In the previous chapters, we've traversed the cutting-edge research and development of AI models and their expansion into new data domains. But in this domain, our journey is characterized not by groundbreaking discoveries but by careful adaptation, stringent fine-tuning, and a tireless pursuit of perfection.

AI applications are far-reaching, spanning sectors as varied as healthcare, finance, manufacturing, retail, and beyond. Yet, these applications are not simply about plug-and-play.

While AI technology may be advancing rapidly, translating this technology into viable, real-world applications demands intricate engineering, deep domain knowledge, and an astute understanding of the specific industry's nuances.

Take a healthcare organization, for example, that is looking to deploy an AI tool for predictive disease modeling. As sophisticated as it may be, the algorithm itself isn't the only factor to consider. Regulatory compliances, privacy laws, ethical considerations, data security protocols, and many other challenges come into play.

The AI tool must be highly accurate and resilient against cyber-attacks, respectful of patient privacy, and compliant with laws such as HIPAA in the United States.

Here's where the technical architects and engineers come into play, the unsung heroes who make the integration of AI into enterprises a reality. Their expertise is pivotal in aligning AI technology with industry-specific requirements and constraints.

Investors who understand the importance of this domain are well-positioned to reap substantial rewards. Companies that demonstrate the capability to integrate AI seamlessly and securely into their business operations can unlock a treasure trove of opportunities. Executing this effectively and efficiently can significantly differentiate a business from its competitors and fuel its growth.

Yet, this terrain isn't just about identifying companies that can aptly apply AI. It also provides a compelling reason for investors to understand how AI may reshape entire industries. In doing so, they can anticipate the industries that are most likely to be disrupted by AI and adjust their investment strategies accordingly.

As the march of AI continues, understanding the landscape of AI applications in industries and specific use cases is essential. It serves as a crucial guide for investors navigating the investment terrain in the AI era. It is the lighthouse in the storm, enabling investors to steer their portfolios towards promising horizons.

The Focused Lens - Tailoring AI for Industry-Specific and Task-Specific Applications

As we deepen our exploration into AI's vast terrain, we see a distinctive trend emerging: the customized adaptation of AI technologies for specific industries and tasks. This trend, more like a strategic necessity, underpins the successful and meaningful deployment of AI systems in the real world. To truly comprehend its importance, let's visualize AI not as a one-size-fits-all solution but as a bespoke suit carefully tailored to each individual or industry's unique needs and requirements.

Broadly, the customization of AI systems occurs along two main dimensions: industry-specific and task-specific. On the one hand, AI systems can be developed to address peculiar challenges and exploit the unique opportunities within different sectors like insurance, travel and hospitality, or government operations. On the other hand, AI systems can also be streamlined to perform specific tasks, such as customer support, sales, product development, human resource management, and more.

To illustrate, consider an AI model designed to assist in the insurance industry. Such a model should understand general language and master the complex terminologies, regulations, and practices peculiar to insurance. Similarly, an AI model developed to streamline HR processes would need to understand job descriptions and hiring practices and have a handle on the nuances of human psychology and cultural sensitivity.

The benefits of such focused AI applications are manifold. One of the primary advantages is the potential for increased reliability and transparency. When an AI system is tailored for a specific task within a specific industry, its learning and application scope is naturally confined. This restriction can significantly reduce the likelihood of bias and hallucinations discussed in previous chapters, thereby increasing reliability and transparency.

In the same vein, task-specific AI systems can also enhance efficiency. For example, an AI system designed to handle customer support in the telecom industry can automate routine queries, freeing up human agents to tackle more complex customer issues. This leads to improved customer experience and operational efficiency.

Understanding the role of customization and adaptation in AI technology is pivotal from an investor's viewpoint. It reveals how value is created by overcoming the challenges unique to each industry and task. The potential of AI to revolutionize enterprise operations is real, and it is unfolding right now. Adopting these tailored AI systems across enterprises will transform how businesses operate and provide immense opportunities for those poised to ride this wave of change. So, as we navigate this path, remember the key to unlocking AI's full potential lies in the details - the careful, precise adaptation to industry and task-specific needs.

Integrated Suites and the Power of Consolidation

As AI applications increase across industries and tasks, a compelling prediction emerges for the future of this dynamic landscape - the advent of integrated AI suites. This evolution is not merely a possibility but a likely inevitability spurred by the complexities of implementation and the underlying network effects that drive the digital world.

The concept of an integrated AI suite stems from the intricacies of data domains and their subsequent applications across various industries and use cases. Today, we see startups in every niche imaginable, from using text data to revolutionize customer service operations to employing image and video data to disrupt marketing and entertainment sectors. Some even delve into audio data, bringing interactive agents or revolutionizing speech training and analysis. While this approach is essential for pioneering the frontier of AI, it is akin to constructing a building by placing bricks in isolation. Each brick is meticulously crafted and indispensable, yet it is only when all the bricks come together, connected in a thoughtful design, that we genuinely appreciate their collective value.

However, each application and data domain brings unique challenges, further amplified when considering enterprise-grade requirements. The customization needs, compliance standards, data security concerns, and implementation complexities all contribute to an intricate, multi-faceted puzzle. Consider an enterprise that employs text-based AI for drafting legal documents, an image-based AI for digital marketing, and an audio-based AI for speech training and analysis. Each of these systems, while powerful in its own right, functions in isolation, limiting the ability to share insights and learn from each other's experiences. Managing data access between various independent suppliers is complex, time-consuming, and risky.

This is where the concept of an integrated AI suite comes into play. Rather than having multiple independent solutions working in silos, an integrated suite would offer services spanning across different data domains and use cases within an enterprise. It paints a picture of a centralized AI-powered system that can handle text, images, videos, audio, and potentially even more exotic forms of data. This system could cater to many applications, from customer support to HR and sales to product development. Herein lies the appeal of integrated AI suites - a single provider offering a holistic AI platform that spans various data domains and use cases. The promise of an integrated AI suite is much like the allure of a well-coordinated orchestra, where each instrument, capable of creating beautiful music independently, contributes to a symphony far more captivating when performed together.

This shift toward integrated AI suites mirrors a pattern we've seen in the past. Reflect to the early days of personal computing. Independent software solutions initially proliferated to address tasks like word processing, spreadsheets, and email. However, over time, integrated suites like Microsoft Office, which unified these functions under a single umbrella, came to dominate. The convenience, efficiency, and seamless interplay of features offered by such integrated suites were too compelling to resist.

A similar trend of consolidation may well be on the horizon for AI. Such integrated AI suites would significantly simplify the enterprise AI landscape, reducing the complexities of data management, ensuring compliance, and improving overall efficiency.

The future of enterprise AI lies in its careful, precise adaptation to industry and task-specific needs, and the winners will be those who can not only adapt but also consolidate. They are poised to shape the symphony of the AI revolution, commanding the lion's share of the inevitable growth and rewards.

So, to our investor readers, heed this prediction: The AI landscape, presently bustling with many startups offering independent solutions, will gradually converge towards a few dominant players providing integrated AI suites. The key is identifying and encouraging companies with the technological prowess, strategic vision, and scalability potential to lead this race toward integrated AI suites. The winners in this space will likely be those who can create a holistic, user-friendly, and effective AI suite capable of meeting tomorrow's enterprises' diverse and complex needs. The race has just begun. As investors, it is crucial to identify these potential winners early on, those capable of harmonizing the cacophony of AI capabilities into a coherent, unified offering.

The Unsung Heroes - Silent Enablers

Intelligent Data Management Systems

While much attention is given to the sparkle and shine of AI technologies and their vast potential applications, a key component of this AI revolution remains uncelebrated. Like an unseen conductor guiding the symphony, it orchestrates the flow of the most valuable commodity in the AI world: data. This unsung hero is the Intelligent Data Management System.

The importance of data in AI cannot be overstated. Data is to AI what fuel is to a car. Without it, the most sophisticated engine is merely an idle piece of machinery. It feeds AI systems, allowing it to learn, adapt, and evolve. Today's AI technologies are fundamentally data-driven. They ingest copious amounts of data to learn, understand, and make predictions. The quality, volume, and relevance of this data directly determine the performance of AI systems. High-quality data is like premium fuel, powering these systems to operate at their best. As we move towards a future where AI permeates every facet of our lives and businesses, the importance of efficient data management grows exponentially.

Until recently, enterprise data was predominantly created and managed with humans in mind. But now, a new player is knocking on our data warehouse doors - our data-hungry AI systems, eager to delve into our data troves and extract insights, patterns, and knowledge. And not just any data; they require well-structured, interconnected, labeled entities that can be easily parsed and understood. This calls for a dramatic shift in managing and curating our data.

Enter Intelligent Data Management Systems, the unsung heroes of the AI era. Their mission is to process, clean, structure, label, and maintain data catering to the unique needs of humans and AI systems. These state-of-the-art refineries take in raw, unrefined data and process it into high-grade fuel for AI systems. These systems transform volumes of disparate, unstructured data into interconnected, refined, and labeled entities, serving as a high-quality feast for our AI systems.

But the role of these systems extends beyond just refinement. They also serve as sentinels, protecting AI systems from the influx of redundant, irrelevant, or potentially harmful data. By curating a continuous stream of relevant, high-quality data, they will ensure AI systems can learn and adapt in real-time, without being overwhelmed or polluted by irrelevant or misleading information, thus making 'online learning' safer and more effective. Furthermore, they ensure the data remains secure, consistent, and accessible, meeting enterprise compliance and governance requirements.

Despite not being the 'fanciest' of technologies, Intelligent Data Management Systems are pivotal in shaping the future of enterprise AI systems. They form the foundation upon which AI technologies can operate optimally.

As an investor, it's easy to get lost in the glitz and glamour of AI's latest and greatest. But remember, even the most sophisticated AI technologies are only as good as the data they're built on.

Companies that prioritize and invest in intelligent data management are likely to be the ones that truly harness the potential of AI.

Unsung Heroes – Silent Enablers

While data is the "fuel" for AI engines, one more player is often overlooked in this domain. "Tools". Imagine an artist, their vibrant vision, many colors, and bristling canvas... but no brushes. Or consider a skilled chef, a panoply of fresh ingredients, a bustling kitchen, but without any knives. In these scenarios, creativity remains stifled, and potential stays untapped. That's precisely where AI finds itself today - a prodigious potential waiting for the right tools to sculpt it into reality. These tools, like brushes to an artist or knives to a chef, are essential enablers that transform AI's vast promise into tangible applications. If AI models are the shining knights of our tale, AI tools are the unsung squires supporting these knights in their conquests. From open-source collaboration platforms to vector databases, AI tools are the invisible hands that grease the wheels of AI innovation. If AI systems are the towering skyscrapers of our technological future, the tools used to develop, deploy, and manage them are the cranes, scaffolding, and safety harnesses that enable their construction. Without these vital instruments, the splendid towers of AI would remain figments of our imagination, trapped on the architect's drafting table.

In AI, these tools are the unsung heroes, the stagehands who make the grand performance possible. They play multiple roles, from drafting the scripts to managing the props to conducting the orchestra to pulling the curtains open for the grand reveal.

Open-source collaboration portals such as Hugging Face emerge as the maestros of this orchestra, serving as the GitHub of the AI world. They provide an inclusive platform that cultivates knowledge sharing, driving rapid advancements in AI. They are the bustling town squares of the AI world, where open-source enthusiasts gather, tweak algorithms, discuss and debate, and improve each other's work, leading to collective progress far beyond any individual's reach. With a vast library of pre-trained models and a forum for discussion, this open-source stage propels the play of AI development.

Vector databases like Pinecone, on the other hand, are the prop masters. They manage and organize the crucial data, or the 'fuel' as we've previously described, that powers the AI system. These tools ensure that our AI 'actors' have the right props at the right time, facilitating a smooth and seamless performance. They manage vast repositories of vectors, ensuring the data, the lifeblood of AI systems, is organized, accessible, and efficient. Akin to a diligent librarian, meticulously categorizing and organizing massive volumes of multi-dimensional data. When the AI knights venture forth in their data quests, tools like Pinecone ensure they find the right information in the right form, at the right time.

But our grand performance also needs a stage manager – something to manage the chaos behind the scenes. And that's where tools for quality control, source control, version management, and continuous integration step in. These tools monitor every performance element, managing the nuts and bolts of AI systems, ensuring that the show always goes on.

Quality control tools are another crucial piece of this puzzle. They serve as the vigilant gatekeepers, tirelessly ensuring that the AI systems' output is reliable, accurate, and unbiased. They help detect glitches early on, saving precious time and resources, and avoiding potential pitfalls.

In the world of AI, change is the only constant. New models, new data, and new techniques emerge every day. This dynamism demands robust source control and version management tools.

They help track changes, manage updates, and foster collaboration, much like a time machine keeping tabs on the evolving landscape of AI development. As blacksmiths forge armor for knights, tools will be developed for quality control, version management, and continuous integration. They will maintain the robustness and integrity of AI models, ensuring that each model iteration is better than the last and that any 'battle scars' from previous deployments are promptly addressed and healed.

Finally, we have the curtain pullers. These tools abstract the complexities of AI, presenting the user with an easy-to-use interface. They are the gateways, opening the world of AI to everyone, not just the tech-savvy.

Imagine an AI 'Lego set' for end-users, where people with no programming experience can create their own chatbot or AI assistant using simple, intuitive graphical interfaces. Like assembling a 3D puzzle, these platforms would allow anyone to build, tweak, and fine-tune their AI while hiding the underlying systems' inherent complexity. The complexity of underlying systems is magically abstracted away, opening up AI for the masses.

We've seen this story unfold before. The past decades have witnessed an explosion of tools supporting software development and cloud migration. Now, it's AI's turn. The coming years will see the emergence of a new generation of tools, the scaffolding upon which the AI systems of the future will be built.

Essentially, these tools are the pillars of the AI revolution. They promise accelerated development and widespread adoption of AI technology, transforming how we interact with our digital world. Investors take note, as the opportunities here are ripe and abundant. The advent of AI isn't just about the grand skyscrapers but also about the tools that enable their construction. Let's remember to value the workhorses alongside the showstoppers.

The Connectors - World of API

Imagine standing on a vast shoreline, gazing at a magnificent ocean. This ocean represents the infinite potential of AI, teeming with waves of innovation and depths of untapped possibilities. Now, imagine a fleet of boats representing an existing product or service. How can we enable these boats to tap into the ocean's vastness, bringing back the valuable treasures of AI innovation to enrich their offerings?

The answer lies in API Integration, the final terrain in our exploration of the AI world. Here, product managers and web developers take the helm, expertly navigating the waters to extract value from the AI ocean and infuse it into their products.

APIs, or Application Programming Interfaces, traverse the boundary between AI's potential and existing products. They allow software applications to access and utilize the functionalities of AI systems, transforming them from mere vessels into intelligent, self-guided boats. Cloud-based AI APIs, in particular, open the gates to advanced AI capabilities for any application, irrespective of its scale, without substantial computational resources.

This is a realm of quick wins and noticeable enhancements. API integrations can swiftly upgrade the user experience of a multitude of products and services that we use daily.

For instance, consider chatbots. APIs for advanced models like GPT-4 can be integrated into chatbots to elevate them from simple query-response systems to more conversational, context-aware, and helpful assistants. They can understand complex queries, provide more accurate responses, and even detect sentiment, leading to a more human-like conversation experience.

From an investor's perspective, API integrations present an attractive avenue for value addition. They offer a way to amplify the impact of AI rapidly across an array of products and services. Investments in companies providing these APIs or those proficiently leveraging them to enhance their offerings can yield significant returns.

Imagine API integrations as a grand ballroom dance where various sectors tango with AI, each creating its unique rhythm and style. Let's peek into this grand dance floor and witness how each industry gracefully waltzes with AI.

In **Healthcare**, AI serves as a personal fitness coach and nutritionist. It whips up customized fitness plans to help users achieve their marathon dreams or merely manage the after-effects of an overly enthusiastic doughnut spree. Need a dietary plan that suits your love for tacos while keeping cholesterol at bay? AI's got your back!

Next, we find **Education** locked in a riveting foxtrot with AI. Imagine a world where studying doesn't feel like a chore. AI creates bespoke study materials perfectly aligned with each student's learning style and pace. Late-night cramming sessions become a thing of the past as AI tutors step in, ready to clarify doubts around the clock.

AI has become the busy reader's best friend in the publishing industry. It can generate crisp, informative summaries for those with overflowing reading lists. Have you ever been stumped about which book to pick up next? Let AI's reviews guide your next literary adventure.

In **Travel and Tourism**, AI is an astute concierge, arranging personalized itineraries that cater to each traveler's quirks. Love museums, despise the tourist crowds, and have a peculiar affinity for tasting exotic cheeses? AI crafts a vacation itinerary that fits you like a glove.

In the world of **Legal** Affairs, AI morphs into a capable paralegal. It can churn out precise legal documents faster than you can say "objection." Those mind-numbingly intricate contracts? A walk in the park for our AI assistant.

Real Estate sees AI as a seasoned property consultant. Need a catchy, detailed description for that quaint mid-century house you're putting on the market? Leave it to AI. Or perhaps you're uncertain about the perfect time to sell? AI's predictive market analysis has got you covered.

Agriculture benefits from AI's wisdom, too. It's like having a seasoned farmer who, armed with centuries of knowledge, suggests the perfect crop rotations and optimized farming techniques. Predicting crop yields based on weather patterns? It's the AI farmer's daily routine.

In this grand dance of API integration, the steps are endless, and the rhythm is ever-evolving. With a bit of imagination, a touch of humor, and a pinch of audacity, there's no telling what AI can do next.

As we close this exploration of the vast AI landscape, it's clear that while the terrain is diverse and the challenges are substantial, the opportunities are truly immense. Each terrain - from groundbreaking innovations to quiet enablers - holds a treasure trove of possibilities. As an investor, discerning where the winds of innovation are blowing, understanding the nuances of each terrain, and spotting the most promising opportunities are the keys to a successful voyage in this AI ocean.

The Investor's Compass - Risk, Reward, Timeline

As an investor, stepping into the realm of Generative AI might feel like embarking on a thrilling yet daunting journey. The terrain is indeed complex with its evolving capabilities, untapped frontiers, looming regulations, ambiguity over data rights and acceptability, and the competitive landscape populated by industry titans and budding startups. But, equipped with the right compass and a keen understanding of the terrain, it's a journey that promises a bountiful treasure of opportunities.

The categorization of AI terrain we've explored provides a solid foundation. It lays out each terrain's challenges, maturity timelines, and impact potential, serving as your compass in the AI landscape. However, a compass alone is not enough; you must also know how to use it effectively.

The key to long-term sustainability in this world, whether you're a startup founder or an investor, is not just understanding the capabilities of AI. It's about choosing a 'terrain of focus' and developing sustained expertise and experience. When assessing potential startups for investment or wading through a flood of pitch decks, it's not just the shiny, exciting technologies that matter. Instead, focus on understanding the company's terrain, depth of expertise, and potential to deliver value within that terrain.

Investing in AI is much like venturing into an uncharted wilderness. With the right compass, understanding of the terrain, and a clear vision of the destination, you'll survive and thrive.

Remember, each terrain requires different sets of expertise. It's like being an explorer. One does not simply decide to explore both the highest mountains and the deepest oceans simultaneously. They require different skills, equipment, and approaches. A mountain climber is not necessarily adept at deep-sea diving. Similarly, attempting to conquer every AI terrain would be overly ambitious and bound to fail, regardless of whether you're a tech behemoth or a budding startup. It might offer tantalizing glimpses of unbounded potential, but it will eventually falter.

As we've discussed before, it's essential to note that technology's 'coolness' factor doesn't guarantee success. Instead, the ability to provide tangible value at the right price—in other words, withstanding market forces—determines the company's success.

As an investor, you should look for companies that have identified their terrain of focus and are developing deep expertise in it. Does the company understand the unique challenges of its chosen terrain? Do they have the right team with expertise to navigate that terrain? Are they focusing on providing tangible value to their customers, or are they getting distracted by the 'coolness' of AI? This doesn't mean you should avoid companies that aim to innovate across terrains.

However, it's crucial to understand whether they have a clear focus to categorize and understand the difference and whether they can deliver in each area. Are they spreading themselves too thin, or do they genuinely have the necessary expertise and resources to innovate across multiple terrains? Evaluating a company's understanding of its chosen terrain, its ability to develop the right expertise, and its focus on providing tangible value are the essential tools you need in your investor toolkit.

In the following sections, we'll explore the specific aspects of each terrain to help you better navigate this journey.

The Foundations Terrain - Playing the Long Game

In the vast landscape of Generative AI, the terrain of core model builders, or 'Foundations' as we have termed it, is akin to the untamed wilderness, where pioneers and explorers gather. Here, the boundaries of possibility are constantly pushed, where the imagination is expanded, and where future AI developments are sown.

Investors have been naturally drawn to this terrain, akin to a gold rush. The intrigue, the hype, the allure of vast untapped potential, and the thrill of being part of an industry rewriting the rules of the digital world all contribute to this appeal. The significant capital investments we've seen in this space are a testament to this attraction.

Players operating in this terrain are:

- The pioneers on the frontiers of possibility.

- Pushing the boundaries of what AI can do.

- Expanding our collective imagination.

- Impacting every downstream terrain.

Yet, this is also a terrain of high-stakes proxy wars between the incumbent tech titans, each vying to conquer this unchartered territory. The dominant strategy so far has been akin to an arms race, with players striving to build bigger and more powerful models.

However, as much as the allure of scale and size has dominated this terrain, a widespread recognition is dawning that 'bigger' might not always be 'better.' There is growing acknowledgment that this strategy of 'might is right' might not be sustainable or productive in the long run.

The pioneers in this terrain are realizing that nurturing this wilderness demands more than mere capital and manpower. Just as you can't reap a bountiful harvest in a week by simply throwing more hands and money at the field, similarly, pushing the frontiers of Generative AI needs more than just capital intensity. It requires patience, a clarity of vision, and a commitment to innovation.

The evolutionary path in this terrain isn't linear. Innovations can spring from myriad sources - established players, open-sourcing some aspects, niche startups with a unique vision, and even university groups pushing the boundaries of academic research.

However, this terrain is also fraught with risks. This terrain is not immune to the laws of creative destruction. The foundations we lay today might crumble tomorrow under the weight of better-performing models, stifling regulations, or shifts in approach. And yet, this constant churn is the essence of this terrain, making it a continually evolving, dynamic space.

As an investor, it's crucial to recognize that investing in the Foundation's terrain is akin to playing the long game. It's about placing bets on the pioneering spirit, the capacity to innovate, and the resilience to navigate challenges. So, while this terrain is alluring, investors must be prepared for the unpredictable journey, armed with the patience to wait for the seeds sown today to bear fruit in the distant future.

The Domain-Data Terrain: Investing in the Application Arena

Navigating the dynamic world of Generative AI, you'll next encounter the terrain of Domain-Data applications. This landscape is less about breaking ground on the uncharted territories of AI fundamentals and more about the ingenious application and adaptation of these breakthroughs to various data domains. In essence, companies operating in this terrain are harnessing the power of foundational AI, tailoring it to work with diverse types of data and addressing a spectrum of user needs. The challenges here are twofold. First, cracking the code of suitable architectures that can handle the magnitude and diversity of data and user requirements. Second, equally critical, is acquiring quality datasets in each data domain.

This terrain's investment risk and opportunity landscape are intertwined with data access. While the current somewhat permissive data access rules might lower the barrier to entry, future developments like stricter copyright regulations or data paywalls could throw a spanner in the works. As an investor, it's crucial to recognize these potential changes in the data access landscape, as they could significantly impact a company's trajectory.

In terms of sectors ripe for exploration, companies have been quick to capitalize on domains where digital technology has a strong foothold, such as software code, video, and audio. However, as technology matures and awareness of its potential expands, we'll see a rise in AI applications in more niche domains that have, to date, seen low AI computer science penetration. Imagine, for instance, the possibilities for generative AI in architecture or fashion design. Investing in such niche domains before they become mainstream could provide an early-mover advantage and open up opportunities for high returns.

Another advantage of the Domain-Data terrain is that technologies here have a shorter path to maturity, given robust market adaptability and acceptance. This makes it a terrain ripe for quicker wins, albeit with its unique set of challenges and potential pitfalls.

Investing in the Terrain of Specific Industry Applications and Use Cases

Welcome to the dynamic and vast terrain of industry-specific applications of generative AI. Here, open-source base models meet specialized data and use cases, creating a patchwork of bespoke applications catering to many sectors and functions. Generative AI applications adapt and conform to niche environments, circumventing the complexities of broad-spectrum AI systems and exhibiting a greater degree of flexibility and utility.

This approach to leveraging AI allows companies to circumvent some complexities associated with large-scale, generalized AI systems. It's like becoming a master gardener of a specific crop rather than attempting to cultivate an entire rainforest.

The unique facet of this terrain is its seemingly fragmented landscape, offering a multitude of niches, each with its unique characteristics, opportunities, and challenges. However, the diverse nature of this terrain makes it hard for one incumbent to dominate across all domains. Consider, for example, fields like Law or medicine. Each of these areas demands an understanding not only of AI technology but also an in-depth knowledge of the field's intricacies, regulatory environment, and user requirements.

The landscape of this terrain is akin to an archipelago of opportunities, each island representing a distinct industry or use case. The potential for AI on each island is vast, yet there's an unmistakable uniqueness in applying AI for each sector. For instance, take the island of Legal AI. While much of the data here is public and thus easier to access, it also demands unparalleled transparency and unbiased output. AI applications here could range from drafting legal documents to predicting lawsuit outcomes.

Similarly, in Medicine, an arena teeming with personal and sensitive data, systems must carefully tread the waters of data privacy and hallucination mitigation. But the potential is immense:

- Personalized healthcare plans.

- Predictive analysis for disease outbreaks.

• Enhancing patient-doctor communication, and much more.

While unique in their demands and offerings, these islands share the need for an in-depth understanding of both the content and the user requirements. This creates a landscape of opportunities that doesn't necessarily favor tech incumbents. Instead, it rewards those who can combine domain-specific knowledge with AI expertise

Within this archipelago, there's a vast island of immense potential and unique challenges—the island of Enterprise AI. While existing "office suite" providers are vying for control of this island, seeking to leverage AI in their products, the enterprise world is not a monolith. A typical enterprise uses different software across various departments, which might not seamlessly share their data due to competitiveness and complexity. This scenario resembles a bustling city where each neighborhood has its unique language, customs, and culture. A solution that works well in one district might not translate effectively in another.

Imagine the enterprise as a bustling city-state within this island. Each department is a district with its distinct language and culture (data and requirements). While each district may have AI tools, there's a glaring need for an "Enterprise AI Operating System," a translator and unifier that can connect these districts, enabling smooth and effective communication and cooperation. These systems would view the "enterprise" as a single, integrated organism, interconnecting various departments like HR, sales, customer support, and documentation. They would navigate the complex web of data domains, including text, images, presentations, audio, video, etc.

Startups that can build this unifying system—providing AI solutions across departments and data domains—stand to make significant strides in the enterprise city-state. Picture a vast network of canals and bridges, connecting the districts and enabling seamless inter-departmental cooperation powered by generative AI. These startups could evolve into the key architects of these infrastructure projects, potentially becoming major players in the Enterprise AI landscape.

Products offering AI solutions targeted at specific enterprise needs might gain rapid traction, similar to the rapid blooming of desert flowers after a rare rain. Still, these blooms are transient, and the desert soon returns to its usual state. The same is true in this terrain of AI. Companies that succeed in the long run will likely be those that can see beyond a single use case. Companies that start with a targeted solution but strategically aim to provide unified enterprise AI solutions are likely to emerge as dominant players in the aftermath of market consolidation

Like the enduring trees in the desert that stand tall amidst the fleeting blooms, these startups would emerge as the dominant players after consolidation and attrition. Their resilience stems from their foresight to provide a comprehensive solution beyond quick fixes, addressing modern enterprises' complex, interconnected needs.

Investors in this terrain need a discerning eye for innovative solutions and visionary companies that can see the forest for the trees. As an investor, your compass in this terrain is a blend of industry understanding, knowledge of AI potential, and an eye for companies with strategic vision and agility. Discing between the transient blooms and the enduring trees can make all the difference.

API Integrations into Existing Products: A Sprint

The last terrain in our exploration of the AI landscape takes us to a realm of immediate gratification - the domain of "API Integrations into Existing Products." Think of this as the bottom of the waterfall, where the fruits of AI technology cascade down and meet the sea of existing applications. In the bountiful orchard of AI innovation, this terrain might seem the least glamorous at first glance. But make no mistake, it is here that we find some of the ripest, low-hanging fruits ready to be harvested. A fertile ground for quick, significant gains, this terrain is teeming with opportunities for innovative companies that can transform and enhance the user experience of existing applications by integrating with cutting-edge AI through APIs.

Imagine a plain old apple tree, robust but unspectacular. The apples it bears are fine but no different from countless others. However, sprinkle in some AI fairy dust (or, in this case, APIs), and the tree transforms. It now grows golden apples, glowing with value added by AI - each one a testimony to the creativity and effectiveness of a simple API integration.

Imagine the expansive and diverse applications: a customer service platform integrating an AI language model to create a hyper-intelligent chatbot capable of deciphering and resolving complex queries, or a fitness app leveraging AI algorithms to generate personalized workout plans that adapt in real time to users' performance. The possibilities are as limitless as the array of existing software applications.

Companies operating in this terrain might not boast a trove of proprietary IP like those in other domains. Still, they offer something equally enticing: accessible and immediate value to the end-user. Using APIs provided by "foundation companies" as a digital bridge, they infuse the power of AI into existing products and services, creating upgrades that feel magical and necessary. They're like innovative chefs creating unique dishes out of common ingredients. While their secret sauce might be a dash of creativity, effective adaptation, or technical prowess, their primary task is identifying areas where AI can add value and deliver it to the end users.

But this quick-win territory can also be akin to a gold rush. Once an application is identified, companies often scramble to capitalize on it. The market can quickly become crowded, with competitors differentiating themselves based on pricing, technological subtleties, or the sheer agility of their development cycles.

In this race, the winners often master the art of rapid expansion. They are the companies that deploy creative business models, forge strategic partnerships, and drive aggressive marketing campaigns. This terrain can be a sprint, where the swiftest, not necessarily the strongest, takes the trophy. These are the swift gazelles of the Serengeti, outrunning the competition to claim the fertile plains of opportunity.

However, investing in this terrain also comes with its share of risks. API integrators are somewhat at the mercy of the core AI providers, who can alter their pricing or tweak their features, which may impact the integrator's offerings.

Think of this as building your house on rented land. It can be a profitable venture, but it's crucial to stay alert to changes in the landscape. Think of it as surfing on the dynamic waves of the AI ocean. Surfers are at the mercy of the wave - a change in the pricing or features of the core AI provider could alter the landscape drastically. Companies must constantly stay alert and agile, ready to ride the wave or risk being swallowed by it.

Investors in this terrain should look for speed, stability, and adaptability. The allure of quick wins can be powerful. Still, the savviest investors would be those looking for companies with the agility to respond to changes, the wisdom to negotiate with AI providers, and the foresight to anticipate and adapt to market trends.

In the relay race of AI evolution, the baton now passes from the long-distance runners to the sprinters. This last stretch may not bring the glory of the foundational advancements or the strategic maneuvering of domain application. Still, it carries the potential for swift and substantial returns. As with any race, it's not just about speed—it's about stamina, strategy, and timing.

As we continue to explore this captivating landscape, let's remember to enjoy the journey. After all, isn't that what all great adventures are about?

Hardware Innovations & The Quantum Frontier

As we conclude our fascinating journey across the various terrains of the AI landscape, let us pause for a moment and turn our gaze beneath our feet. There, we find an essential foundation, a realm less talked about but no less significant - the kingdom of hardware.

In all its awe-inspiring complexity, AI's magic is predicated upon the computational power of GPUs and other dedicated hardware. These devices are the bedrock, the unsung heroes of the AI ecosystem. The towers of code, the sprawling databases, the ever-evolving models—all stand tall on this firm foundation.

As the demand for more sophisticated and efficient AI systems grows, we anticipate the birth of new hardware tailored specifically to cater to the demands of Generative AI workloads. This could be in the form of custom accelerators, optimized memory systems, or other bespoke designs. These improvements are confined to colossal cloud data centers. They will penetrate the burgeoning domain of edge applications, where AI is pushing boundaries closer to where data is being generated.

Yet, on the horizon, we see another frontier emerging, one that holds immense potential: the world of quantum computing. Like a distant lighthouse, it beckons, promising a revolution that could change the face of AI and the entire field of computing. Progress might be steady rather than rapid, but its potential is astounding. Although still in its nascent stage, it's steadily pulsating, promising to leapfrog our processing capabilities into a new era. If today's hardware is a galloping horse, quantum computing could be the Pegasus, ready to take flight.

The AI story is being written: one line of code, one silicon chip, and one quantum bit at a time. As we journey into the heart of this brave new world, let's remember to honor both the lion's roar and the elephant's strength. In this symbiosis of software and hardware, this dance between the seen and the unseen, the tangible and the intangible, that the future of AI truly lies.

As we conclude this narrative, remember that while we've primarily viewed this spectacle through the investor's lens, the insights we've gleaned apply equally to the other side of the coin— the startups and entrepreneurs eager to ride this AI wave. The opportunities, challenges, and strategies are as relevant to those creating as those investing.

So, here's to the visionaries, the builders, the risk-takers, and the dreamers.

8

PRUNING THE DISRUPTION: GOVERNMENTS

Unfurling like a wave of electric vitality, Artificial Intelligence (AI) is steadily becoming the ubiquitous lifeblood of our modern world. It courses through the veins of our global society, stimulating economic growth, fostering innovation, and reshaping our individual and collective futures. It's everywhere - in the devices we carry in our pockets, the cars we drive, the ways we entertain ourselves, and the silent networks connecting our increasingly digital lives. No longer confined to science fiction, AI now permeates every facet of our existence—healthcare, education, finance, environmental conservation, and everything else. It is a revolution, a cascade of transformational potential that transcends borders, industries, and social strata. And it is in this context of accelerating change that the role of government in managing the AI revolution becomes a matter of profound importance.

It's an era of immense possibilities in which innovation and pioneering are not only encouraged but are often the linchpins of progress.

This great revolution that promises a sea of possibilities also casts a shadow of uncertainty. If left unchecked, AI could engender social upheaval, breed disparity, and compromise the fabric of our social norms and ethics. It is a two-faced Janus, capable of propelling our civilization towards unimaginable heights of prosperity while equally bearing the potential to plunge us into a chasm of social unrest and disparity. AI's complex, dichotomous nature necessitates a measured, deliberate approach to its evolution. This approach respects this technology's potential while mitigating its possible adverse effects. Here, the role of governments becomes not just pertinent but indispensable. Harnessing its power while mitigating its perils requires a measured, nuanced, and proactive approach.

As the architects of our shared social contract, governments have traditionally been seen as observers rather than active technological participants. But in the world of AI, this paradigm is shifting. Governments now have a crucial role, a delicate balance to strike.

While not generally involved in the intricacies of technological advancement, they are presented with a formidable challenge and a compelling duty. Their toolbox for shaping the AI landscape ranges from incentives that promote beneficial innovations to regulatory brakes that temper the headlong rush of unrestrained development. The governmental hand has the power to tip the scales, facilitating a sustainable trajectory of AI development that prioritizes the welfare of society at large.

In the past, adopting new technologies has often followed a predictable pattern. Once they transition from theoretical research into tangible products and mainstream usage, governments begin the catch-up game. Legislation and regulation lumber after the retreating of progress, striving to ensure the larger social good in the face of rapid change. But the AI revolution, with its rapid, relentless, and pervasive march, threatens to outpace the conventional rhythm of governance. In this relentless race against technological progress, the finish line keeps moving. The challenge for governments is that technology, particularly AI, is evolving at a breakneck pace that far outstrips the traditional rhythms of bureaucracy and legislation.

This dissonance between technological progress and regulatory adaptation can result in a regulatory void. In this vacuum, technology advances without adequate checks and balances. This lag is not a new phenomenon. We saw similar delays in developing and implementing data privacy laws following the explosion of user-tracking technology, leaving consumers vulnerable to exploitation.

Regulation is a highly nuanced and complex task in the context of AI. The balance between societal protection and innovation is precarious, like walking on a tightrope. Regulating for social benefit and citizens' rights without throttling the spirit of innovation requires a deep understanding of the technology in question, a comprehensive grasp of its implications, and a visionary foresight into its future trajectory.

Enact legislation too strictly, and the oppressive grip of bureaucracy risks suffocating innovation, allowing other nations to surge ahead. Yet, allowing a laissez-faire approach can lead to a dystopian landscape where AI is misused, manipulated, or monopolized. The multi-stakeholder and multi-dimensional nature of the task makes building consensus, drafting, and implementing laws a Herculean task.

The inherently borderless nature of AI further magnifies this challenge. We can no longer view technology through isolated domains or geographical boundaries. With the internet's ever-expanding reach, AI applications permeate every corner of the globe, magnifying the urgency of the situation.

Politicians, bureaucrats, and policymakers must become "AI-literate." They must delve deep into AI, understanding its nuances, potential, and impacts. Governments urgently need to equip themselves with the knowledge to navigate this uncharted terrain. Understanding and regulating AI is no longer confined to the tech-savvy elites but is necessary for all those involved in shaping our collective futures. The role of government in managing the AI revolution will be as integral as the technology itself.

The dawning of the AI era presents governments worldwide with a complex set of responsibilities and opportunities. While these span a vast range of areas, they can be grouped into three pivotal domains: utilization for national interest, regulation for public wellbeing, and thought leadership coupled with establishing centers of learning and economic development. Each of these facets is a cornerstone for constructing a robust framework for the sensible, effective management of the AI revolution.

1. **Harnessing AI for National Interests:** The first challenge laid at the doorstep of governments is to utilize AI effectively in pursuing national interests. From bolstering defense capabilities to optimizing public services, the canvas for AI applications is as diverse as it is vast. In realms such as national security, AI can provide unprecedented surveillance capabilities, intelligent threat detection, and robust cyber defense mechanisms. Meanwhile, the public sector can benefit from increased efficiency, improved decision-making, and heightened levels of transparency, making the dream of smart, responsive governance a tangible reality. However, the task is not as simple as it may sound. Governments must tread a careful path to ensure that the adoption of AI doesn't unwittingly compromise the very interests they seek to protect. This could include data privacy issues, the potential for AI-enhanced surveillance states, and the risk of autonomous weapons in the military. Harnessing AI for national interests calls for a sophisticated

understanding of the technology, a clear-eyed appraisal of its potential risks, and the wisdom to employ it effectively without undermining the values and principles a nation stands for.

2. **Regulating AI for Public Wellbeing:** The second key responsibility of governments lies in safeguarding the public's wellbeing in the face of AI-induced transformations. As AI permeates everyday life, from shaping our digital experiences to influencing our job prospects, its impact on public wellbeing becomes increasingly significant. Governments need to erect regulatory bulwarks that ensure the AI revolution uplifts rather than undermines the lives of their citizens. Crafting such regulations, however, is an act of balancing on a high wire. On the one hand, governments must protect individual privacy, prevent algorithmic bias, and mitigate the risk of job displacement due to automation. On the other hand, they need to ensure that these regulations do not stifle the innovation that could lead to societal advancement. This requires the kind of informed, flexible, and proactive governance that can foresee potential issues and respond swiftly to unforeseen consequences.

3. **Providing Thought Leadership and Fostering Centers of Learning:** The final puzzle piece is the government's role in providing thought leadership and fostering learning centers to harness AI's economic potential. As the vanguards of societal progression, governments need to take the helm in defining the narrative around AI, shaping its perception from a potential threat to an ally of progress. Beyond mere rhetoric, this involves nurturing an ecosystem where AI research and development can thrive. This could range from funding educational programs that train the workforce of the future to establishing research institutes that push the boundaries of AI innovation. It could also

create an entrepreneurial climate where AI-based startups can flourish and contribute to economic development. This responsibility is the most forward-looking, requiring a commitment to long-term investment and the courage to champion a misunderstood technology. In fulfilling this role, governments have the opportunity to foster an environment that leverages the full potential of AI while ensuring that its benefits are widely distributed across society.

The following sections will delve deeper into this fascinating intersection of AI and governance. We will dissect the challenges, celebrate the potential, and imagine a future where AI and governance work together, striving toward a fair, safe, and prosperous society. As we journey through these responsibilities, we'll discover that each one is not an isolated silo but part of an interconnected triad. The effective management of the AI revolution demands a comprehensive, integrative approach that harmonizes these three roles into a cohesive, holistic strategy.

Sentinel of National Interests

As we traverse further into the digital age's labyrinth, warfare's boundaries have morphed beyond the conventional battlegrounds. Today, the silent hum of servers and the invisible weave of networks bear the brunt of attacks threatening to destabilize nations. The frontlines are no longer guarded by soldiers but by complex lines of code and vigilant algorithms operating on the battlefield of cyberspace.

In the crosshairs of these sophisticated cyber assaults lie the vital arteries of national infrastructure—utilities, transportation networks, communication systems—all interconnected and increasingly vulnerable. Unfortunately, the digital weaponry that threatens these critical lifelines evolves staggeringly, powered by the technology we strive to protect ourselves with—Artificial Intelligence.

The double-edged sword of AI stands at the forefront of this digital combat. On one hand, it equips malevolent actors with formidable tools to mount cyber-attacks of unprecedented sophistication. Automated hacking tools, programmed with the cunning of AI, tirelessly probe and exploit system vulnerabilities, far outpacing the capacity of human counteraction. AI empowers phishing emails to mimic genuine communications with alarming accuracy, making them deadly digital trojan horses. The specter of AI-powered ransomware looms large, capable of bringing essential services to their knees, eroding public trust, and sparking widespread chaos.

Consider the explosive impact of AI-driven malware striking the heart of our digital lifeblood—data centers, cloud infrastructure, and internet providers. The fallout would not merely be limited to data theft, as crippling as that would be. Imagine the paralyzing consequences as businesses halt: transactions freeze, hospital records vanish, and everyday conveniences like food delivery systems stutter. An interconnected digital landscape, though brimming with convenience, can spiral into a chaotic dystopia when brought to a sudden standstill by a well-orchestrated cyber-attack.

Compounding the threat of these sophisticated cyber assaults is their uncanny ability to evolve, much like a biological virus evading an immune response. Traditional cybersecurity measures often rely on recognizing patterns from previous attacks, an increasingly ineffective strategy against AI's relentless adaptability. Outsmarting this new breed of cyber threats requires a vigilant sentinel that matches the cunning of its adversaries and outwits them at every twist and turn.

Here, the second face of the AI, Janus, comes into play—the invaluable role of AI as the protector. Governments worldwide must urgently prioritize developing and deploying AI-powered defense mechanisms to match the evolving threats. This is a task that goes beyond simply implementing new technology. It requires an intricate understanding of AI, its potential, and the diverse threats it could be twisted to unleash.

The very strengths of AI—its speed, precision, and capacity for complex problem-solving—can be harnessed to enhance national cybersecurity. Generative AI systems can shoulder routine but intricate tasks, liberating skilled cyber personnel to tackle the ever-changing landscape of threats. By leveraging AI's potential, we can build reactive defense systems that proactively anticipate, adapt, and neutralize threats, keeping us one step ahead in the ceaseless game of digital cat and mouse.

To attain this level of sophisticated defense, a commitment to extensive research, infrastructure development, and skill enhancement is imperative. This may demand significant investments, but the price of inaction—leaving our digital fortresses unarmed against the relentless onslaught of AI-powered threats—could be catastrophic. As our world tightens its embrace of the digital, safeguarding national infrastructure is not a choice but a non-negotiable imperative.

Let's now turn our attention to social media. Misinformation and malicious activities are rampant, causing havoc and inciting unnecessary tensions. Enter AI. AI can flag harmful content or identify suspicious patterns by analyzing millions of posts and trends. A surge in posts with specific dangerous contexts, an influx of images with hidden illicit messages, or the rapid spreading of fake news - AI can detect all this and more, helping moderators act swiftly.

An important caution here is that while AI can serve as a watchful sentinel, it cannot and should not replace human judgment. Especially with complex ethical decisions and nuances of human communication, the AI's role should be to aid, not overrule, human decision-makers.

Preserving Legacy: National Knowledge Management

"Those who cannot remember the past are condemned to repeat it." These wise words by philosopher George Santayana underscore the value of knowledge, especially when it relates to our cultural, historical, and traditional heritage.

The tapestry of a nation's history, woven with the threads of milestones, cultural narratives, and traditional wisdom, is a testament to its past and a beacon guiding its future. However, this repository of national knowledge faces an existential threat in the digital age, where the rapid inundation of new information can overshadow the voices of our past.

But what if we could harness the power of artificial intelligence (AI) to breathe new life into these treasures? What if we could secure our legacy not merely in dusty, forgotten corners of antiquated libraries but in the dynamic, interactive world of the digital age? This is where AI could be the game-changer, not just as a vessel carrying our heritage into the future but as a bridge that connects our past with our present and future generations.

Imagine a world where our historical records, cultural narratives, and traditional wisdom are digitized and democratized. Where anyone, at any time, can access these treasures at the click of a button. And we're not talking about static, inert documents. Instead, envisage an interactive, AI-powered narrative that can engage, enthrall, and educate. A narrative that speaks, listens, and converses, a narrative that brings history alive and makes our heritage tangible.

The creation of national AI repositories could be our bulwark against cultural amnesia. These repositories would not just preserve our national knowledge. They would revitalize it. The voices of our ancestors would echo in our daily lives, their wisdom guiding our actions, their experiences shaping our decisions.

These repositories would not be mere passive storehouses but dynamic knowledge ecosystems. Large language models could

transform how we interact with our heritage, allowing us to converse with our past rather than merely observe it. Searching for information in vast, disconnected records could be a thing of the past. Instead, we could engage in an intuitive, chat-based system that unveils connections between disparate texts, bringing coherence to the cacophony of our past.

Furthermore, these repositories could empower AI to offer insights and solutions deeply rooted in our unique cultural contexts. Armed with a diverse and comprehensive dataset, AI models could draw from the vast pool of our heritage to inform our present and guide our future. Our legacy would cease to be a passive testament of our past but become an active participant in our present and a visionary architect of our future.

Imagine the power of visualizing history as it was described in ancient texts, not through the distorted lens of centuries of interpretation but through the clear prism of the original narratives. We would not be looking at history but looking through history, an invaluable perspective that would enrich our understanding of our cultural identity.

Preserving our national knowledge in the AI era is a challenge that transcends technological boundaries. It straddles the realms of culture, history, tradition, and technology. It is a call to action to ensure our unique identities are not eclipsed in the digital age. It is a quest to harness the transformative power of AI to secure our legacy and carry it into the future.

Let's transport ourselves to this brave new world where AI breathes life into our cultural heritage.

Imagine a high school student, Maya, working on a project about her country's traditional folklore. Instead of rummaging through a library or aimlessly surfing the internet, she logs into the national AI repository. Unlike chatting with a friend, she converses with the AI model. "Tell me about the traditional folklore of our country," she types. Having learned from the digitized cultural narratives stored in the repository, the AI starts weaving stories of old, describing legendary heroes, ancient customs, and the timeless wisdom hidden in these tales.

Impressed, Maya asks, "Can you show me what the ancient city from the folklore looked like?" With the integration of Generative AI models, the AI produces an image, a vivid representation of the ancient city based on descriptions found in historical texts, immersing Maya into the past. She doesn't just read about her culture's folklore. She experiences it, guided by AI every step of the way.

Now, consider a policy-maker, David, grappling with a complex societal issue. He needs insights that only a deep understanding of his nation's history can provide. Instead of consulting an array of experts or delving into hundreds of books, he turns to the national AI repository. He asks the AI, "What have been our traditional approaches to dealing with similar issues?" Using its vast knowledge derived from the repository, the AI outlines a detailed analysis of historical responses, highlighting successes, failures, and the reasoning behind each approach. David uses these insights to inform his current policy, thus leveraging historical wisdom to navigate contemporary challenges.

These examples offer just a glimpse into AI's transformative power in preserving and propagating our cultural heritage. They underscore AI's potential not just as a tool for preservation but also as a platform for engagement, education, and enlightenment. By marrying our past with AI, we're safeguarding our heritage and gifting it a new lease of life.

Promoting Accessibility of Legal Systems

Throughout history, the scales of justice have often been tipped against the layperson. Legal complexity, bureaucratic red tape, and policy obfuscation have acted as an invisible wall, keeping the common citizen from fully understanding and engaging with their rights and responsibilities. It is a world where the letter of the law is often buried under layers of legal jargon, where policy updates hide behind the veil of archaic language. But what if we could dismantle this wall? What if we could bring the mountain of the law to the people? Enter the era of Artificial Intelligence, where the dynamics of legal accessibility might be on the cusp of a revolution.

Imagine a future where an AI-powered assistant is your guide through the labyrinth of legal statutes, a beacon in the bureaucratic fog, and a translator who deciphers policy updates into your language. It is not merely about understanding the law but making it work for you. This is where AI can be a catalyst for change, breaking down the barriers that have long kept legal knowledge confined within the ivory towers of legal scholars.

By leveraging the capabilities of Natural Language Processing (NLP), AI can play a crucial role in simplifying legal language and translating cryptic legal terminology into understandable terms. Imagine a world where citizens can question a smart assistant about a specific law or policy and receive an explanation in layman's terms. In this world, the law speaks to you, not to you. Beyond interpretation, AI can potentially revolutionize how we seek legal help. Legal aid, often expensive and time-consuming, could be made more efficient and affordable. AI could guide individuals to the relevant legal resources, help identify pertinent laws and policies, and even automate legal proceedings. AI could act as your legal advisor, a constant companion in your journey through the legal landscape.

However, the transformative power of AI extends beyond just aiding in understanding and seeking legal help. It also can revolutionize how laws, policies, and changes are disseminated to the general public. Instead of diving into the depths of complex documents and webpages, people could interact with AI tools that can distill information based on their context and understanding. Imagine an AI system that can explain the same legal concepts to individuals at different levels of sophistication using different language styles. It would be a personalized legal guide tailored to the unique needs of every citizen.

It's not just about bringing legal systems to the people; it's about empowering them to navigate these systems themselves. It is, in essence, a move towards truly democratizing our access to legal systems, and AI stands at the forefront of this exciting paradigm shift.

As we delve deeper into this topic, we will explore the cutting-edge technologies driving this revolution, discuss the inherent challenges, celebrate the victories, and chart a roadmap for the journey ahead. As we stand at the crossroads of this transformative era, we must remember that the true power of AI lies not just in its technological prowess but in its potential to empower individuals and societies. Let's explore how we can harness this power to ensure that the law, in its letter and spirit, truly belongs to the people.

Let's envision this scenario:

Meet Leila, a 65-year-old retiree who has recently moved to a new city to live closer to her grandchildren. While tech-savvy for

---

her age, Leila is new to the city's laws and regulations. One day, she received notice that her newly bought property might infringe on local zoning laws. Overwhelmed and uncertain, she turns to her smart speaker and says, "Help me understand this zoning law notice."

The AI assistant, powered by an advanced language model, reads through the notice document. In simple, clear language, it then explains to Leila what the notice means. It's about a particular local zoning law that limits the height of garden fences, which Leila's currently exceeds. The AI assistant then guides her through applying for a variance permit, explaining the steps involved and even helping her draft the necessary application letter.

For Leila, what could have been a stressful and potentially costly encounter with the legal system becomes a manageable task, all thanks to AI.

Let's consider another example. John, a small business owner, struggles to understand the implications of a newly introduced tax law on his business. He opens an app on his phone which hosts an AI legal assistant. He types in his question about the new law, and the AI, programmed with the latest updates and regulations, presents the information in a way that John can easily understand. It explains how the new tax law affects his business and even offers suggestions for optimizing his finances.

Not only does this save John a trip to his (quite expensive) tax consultant, but it also empowers him with knowledge and the confidence to make informed decisions about his business. This AI tool could even send alerts when there are new laws or changes to existing ones that could impact John's business.

These are just a few examples illustrating the enormous potential of AI in democratizing access to legal systems and transforming how we interact with laws and policies. It's a world where everyone has a personalized legal guide at their fingertips, helping to create a more informed and empowered citizenry.

AI LITERACY IMPERATIVE FOR POLICY MAKER

SECURING NATIONAL INTERESTS

1. Cyber security & protecting critical infrastructure

2. National knowledge management

3. Making law accessible

4. Securing Economic potential of public data

PUBLIC WELLBEING

1. Tackling Misinformation

2. Effective governance operations

3. Tackling rapid job-loss with advancing AI

4. Data Use regulations

THOUGHT LEADERSHIP

1. Channelizing AI development

2. Promoting widespread AI Literacy

3. Ministry of AI

Consider this real-world illustration to understand the depth and scope of AI's contributions to tax fraud detection and social media analysis.

Envision a small country plagued by rampant tax evasion. Billions are lost every year due to fraud, undermining public services and increasing the burden on honest taxpayers. The government has decided to employ AI to tackle this issue. The AI is trained on years of tax records, learning to identify patterns that might indicate fraudulent activity. It cross-references multiple data sources, flags suspicious anomalies, and alerts human auditors for further investigation.

With each case, the AI learns, and its ability to detect potential fraud improves. The results are transformative—billions are reclaimed in lost revenue, and the scale of tax evasion shrinks dramatically. The message is clear to the public—tax evasion will not go unnoticed, fostering a culture of compliance and equity.

This, however, should not mean that citizens' privacy is infringed upon. Strict measures need to be in place to ensure that AI doesn't misuse or mishandle personal information, using only what is needed and discarding what isn't.

And there's more. These are just two applications; AI's potential to improve governance is vast and largely untapped.

AI Embedded Governance

Let's now turn our attention to the internal workings of government institutions. The application of AI isn't merely an external pursuit aimed at public services and protections. It holds substantial potential to enhance the effectiveness of operations and inter-departmental communications within the government machinery.

Consider the vast amount of data that government agencies generate and handle daily. This includes bureaucratic documentation, citizen service requests, regulatory filings, and countless other forms of information. Traditional manual processing of this data can be time-consuming and fraught with inefficiencies. AI can revolutionize government operations by automating mundane tasks and streamlining processes, thus freeing up human resources for more strategic roles.

AI-based systems can help sort, categorize, and prioritize incoming requests and workloads, enabling government staff to manage tasks more effectively. Natural Language Processing (NLP) tools can parse and interpret written communication and automate tasks like summarizing reports, preparing meeting minutes, or tracking action items. Another critical application lies in enhancing cross-departmental communication. Traditionally, silos within government bodies can impede information flow, leading to duplicated efforts and missed opportunities for collaboration.

By applying AI to internal communication and data-sharing, we can break down these silos. Machine learning models could predict which departments need to collaborate based on the nature of their work or flag overlapping initiatives that could be merged for greater efficiency.

Imagine a scenario where a country has developed robust internal AI capacity. This isn't merely a repository of AI tools but a deep, institutional understanding of AI's potential and limitations. From the highest echelons of decision-making to the frontline workers, this AI literacy could enable a more effective, proactive, and resilient government body.

This internal AI capacity would serve as a foundation that can be leveraged across various arms of the government, military, and police. With a central AI competence center, knowledge transfer, best practices, and successful AI implementations could occur smoothly, ensuring all departments can access the best available AI resources. This could lead to more cohesive strategies, informed decision-making, and efficient resource utilization.

AI has immense potential to transform how governments operate, internally and externally. By enhancing internal efficiencies, breaking down silos, and building robust AI capacity, governments can position themselves to navigate the challenges of the 21st century more effectively.

Let's consider some practical examples.

Streamlining Operations

Imagine a government department that handles infrastructure projects. It receives thousands of documents daily, including proposals, bids, contracts, reports, compliance certifications, etc. Processing all these manually is time-consuming and prone to human error. Here's where AI comes in.

An AI-powered document management system could streamline this process. As soon as a document arrives, the system classifies it (is it a bid, a contract, or a report) and routes it to the appropriate official or team. The system could also extract key information from the document — for example, the project's budget from a proposal or the completion date from a report. This extracted information could be automatically summarized and presented to decision-makers, helping them quickly grasp the document's essence without reading it.

Moreover, if the document needs certain action – say, it's a proposal that needs approval or a compliance certification that requires a review – the AI system could automatically schedule this action, ensuring nothing falls through the cracks.

Or consider the case where a national environmental agency is tasked with reviewing and responding to thousands of public submissions about local environmental concerns each year. Traditionally, this involved a team of employees manually reading and categorizing each submission, which could take weeks.

However, an AI system automatically categorizes these submissions based on their content upon receipt. The AI system flags urgent matters, assigns them to the relevant personnel, and sends notifications for immediate action. The system can also generate a summary of the issue for the personnel to review, significantly reducing the processing time and allowing the agency to respond to public concerns swiftly and efficiently.

Enhancing Cross-Departmental Communication

Let's think about a complex national problem, like combating climate change. This task involves multiple departments — Environment, Energy, Transportation, Agriculture, and others. Traditionally, these departments might work in silos, leading to fragmented efforts.

Here's where AI could enhance cross-departmental communication. An AI tool could detect overlaps or complementary efforts by analyzing each department's various projects, initiatives, and communications. For instance, if the Environment Department is working on a reforestation initiative, and the Agriculture Department is planning a program that would involve cutting down trees in the same area, the AI tool could flag this conflict and suggest collaboration.

The AI could also identify gaps where collaboration is needed but not happening. For example, it could recognize that the Energy Department's new renewable energy policy might impact transportation trends and suggest that the Energy and Transportation Departments coordinate.

Let's consider a situation where a health department is running a campaign to promote healthy eating habits while the education department is developing a new curriculum that includes nutrition education. Traditionally, these two departments might work in isolation, duplicating efforts. However, an AI system could identify the overlapping objectives and suggest a collaborative approach, potentially creating a more comprehensive and impactful program.

These are just a couple of examples that illustrate how AI could streamline operations and enhance communication within governments. By automating routine tasks and promoting collaboration, AI has the potential to make governments more efficient, effective, and responsive. It's an exciting prospect, one that could shape the governments of the future.

Regulation of Public Data Use

As AI continues to evolve, data becomes its lifeblood. Every interaction we have online, every transaction we make, and even our geolocation information serve as valuable data that can feed into AI models. As a result, the balance between data generation, data privacy, and the need for data in AI development has become a pressing concern.

Governments are vital in formulating and enforcing data regulations that strike this balance. Privacy laws must ensure that personal data is not harvested without informed consent and is used only for the purpose for which it was collected. Governments can also regulate how data is stored and how long it is kept to minimize potential misuse.

Governments can also encourage the development of privacy-preserving AI technologies such as differential privacy and federated learning. These technologies can allow AI to learn from data without compromising individual privacy.

At the same time, governments can promote open data initiatives, releasing non-personal data in areas such as weather, transportation, and public health. This can provide valuable fuel for AI development while avoiding the risks associated with personal data.

AI's potential for commercial exploitation is both enormous and fraught with risk. Companies can use AI to target consumers with hyper-personalized advertising, potentially infringing on privacy and manipulating consumer behavior. AI can also be used to set prices dynamically, leading to potential unfair pricing practices.

To prevent such misuse, governments can establish and enforce strict regulations. For example, they can require businesses to be transparent about AI use, including advertising and pricing. They can also introduce "right to explanation" laws, which require companies to explain AI decisions that affect consumers.

The government can also ensure businesses don't use AI to reinforce or amplify existing inequities. For example, if a company's AI model is trained on biased data, it could lead to discriminatory outcomes. Governments can prevent this by requiring businesses to conduct bias audits of their AI models.

Regulating AI in the commercial sector requires a deep understanding of technology and its societal impacts. Governments have an important role in this, but they must also ensure that regulation doesn't stifle innovation. It's a fine line to walk, requiring careful, informed decision-making.

In this digital age, the notion of truth, or our collective agreement on what constitutes truth, is under threat. The proliferation of disinformation, misinformation, and outright lies, often weaponized and disseminated via social media platforms, has taken the form of a deluge that threatens to sweep away the firm ground of shared reality.

Amidst this chaotic landscape, governments are responsible for ensuring the integrity of public discourse. In essence, the battleground has shifted from the physical realm to the cyber sphere, where governments must act decisively to protect the democratic fabric of society.

Governments have traditionally been the custodians of public discourse and the architects of national narratives. However, the landscape has significantly changed with the advent of social media and its unprecedented reach. Now, virtually anyone can shape the narrative, making it increasingly complex to distinguish between fact and fiction. This dynamic has significant consequences, as distorted narratives can manipulate public opinion, sway elections, and even incite violence. The need for governments to regulate the dissemination of information has never been more crucial.

But how can this be achieved? One approach is through monitoring and fact-checking. Governments could collaborate with AI research institutions and social media companies to develop sophisticated AI systems capable of detecting and flagging false or misleading content. Additionally, public investments can be directed towards enhancing digital literacy, which would equip citizens with the necessary skills to discern truth from falsehood in the digital arena.

Regulating online platforms is another essential strategy. Governments must hold these platforms accountable for the content they host, encouraging them to actively moderate and fact-check the information they disseminate. One possible measure could be implementing strict penalties for platforms that fail to remove harmful content.

This, however, is a delicate task. On the one hand, strict regulation is necessary to curb the spread of disinformation. Conversely, not infringing upon the fundamental right to freedom of speech is crucial. Striking a balance between these two imperatives is a task that governments will need to navigate with care and precision.

Moreover, governments must maintain high transparency and integrity standards in communication to set the right precedent. Honesty from the government can reinforce trust among citizens, making them less likely to fall prey to misinformation campaigns. It is not enough to point out falsehoods; governments must also ensure they are credible sources of truth.

In conclusion, the battle against the deluge of disinformation and misinformation is a challenge of paramount importance for governments in the AI era. Through proactive measures such as monitoring, fact-checking, regulation, education, and maintaining transparency in communication, governments can provide a bulwark against these harmful tactics. This task is fraught with complexities and potential pitfalls, but it is one that governments cannot afford to ignore, for the stakes are high concerning nothing less than the integrity of our democratic societies.

Combating the Specter of Job Loss

The dawning era of artificial intelligence brings many complex challenges that governments worldwide must confront head-on. Of these, one of the most daunting is the specter of job displacement due to the rapid advancements in AI technology. The issue is akin to an approaching storm, quietly gathering momentum on the horizon. At the same time, we continue our daily lives, blissfully oblivious to the potential disruption. This isn't a distant possibility; it's a reality that we are starting to grapple with today. In social welfare countries, where unemployment could strain the fabric of society and fuel social unrest, the need for effective solutions is particularly urgent.

As AI technology continues to advance and permeate every sector of our economy, from healthcare to manufacturing, the fear that machines could replace human labor on a massive scale is no longer the stuff of science fiction. According to estimates, millions of jobs across various sectors are at risk of automation within the next few decades. While AI is expected to create new jobs, these jobs may require skills that the current workforce does not possess.

The narrative often heard is that technology creates more jobs than it destroys. That may be true in the long term. History certainly validates this view, but what happens in the transition phase? Rapid AI advancement could create a 'gap period' where job losses outpace creation. People aren't machines; they can't be reprogrammed overnight. The process of reskilling takes time, effort, and resources. But jobs are disappearing faster than people can retrain, and therein lies the problem.

Faced with this impending storm, the "ostrich approach" of burying our heads in the sand is simply not an option. Instead, governments must adopt a proactive stance, addressing the challenge of job displacement head-on. This means acknowledging the reality of job displacement and taking steps to prepare for it.

As AI continues to shape our world increasingly profoundly, the government has an essential role in educating the public about its impacts. Clear, truthful, and nuanced communication about AI's capabilities, limitations, and implications is crucial. The public deserves to know the reality of the AI evolution, unfiltered by the hype often generated by enterprise narratives showcasing their competitiveness.

We must break down AI into two major categories: Public AI and Enterprise AI. Each has unique opportunities and challenges that the general public must understand.

Public AI systems are those we interact with regularly. The social media algorithms determine the news we see, the smart assistants on our phones, the recommendation systems on streaming platforms, and more. While these systems have made our lives more convenient, they come with challenges we must acknowledge.

One key challenge is the dissemination of disinformation. AI algorithms that keep users engaged can unintentionally propagate false or misleading information, leading to wide-scale misinformation campaigns with real-world implications. Another challenge is AI "hallucination," where AI makes incorrect predictions or assumptions based on its learning. Remembering that AI doesn't 'understand' as humans do is vital. Their decisions are based on patterns they've learned from data, which can sometimes lead to incorrect or biased outcomes.

Enterprise AI, on the other hand, is employed within companies to increase efficiency, drive innovation, and boost competitiveness. These systems can automate tasks, analyze massive datasets quickly, and make predictions that assist strategic decision-making. However, as these systems become increasingly sophisticated, they pose a significant challenge: job displacement due to automation.

Education about both these AI categories should be a national priority. Governments could create educational programs, both online and offline, to raise awareness about AI's workings, the ethical considerations surrounding its use, and its societal implications.

Engaging experts from academia, industry, and non-profit organizations to craft these programs would ensure a balanced perspective. These educational initiatives could use simple, relatable language and real-life examples to explain complex AI concepts, making them accessible to everyone, irrespective of their technological proficiency.

Informative campaigns on social media, TV, and radio, along with community-based workshops and seminars, could disseminate this knowledge to a wide audience. Introducing basic AI education in schools could ensure future generations become informed digital citizens.

An informed public is an empowered public. By educating citizens about AI, we alleviate their fears and help them harness its benefits while remaining alert to its potential pitfalls.

Thought Leadership

Governments, with their unique position and mandate, can adopt a visionary approach to AI, navigating its myriad challenges and opportunities with a perspective set firmly on the horizon of the future. While the dynamic forces of the market drive technological innovation and economic profitability, governments play a critical role in ensuring equitable benefits for all citizens, especially those derived indirectly from these advancements.

In this context, governments can focus on leveraging AI to build and augment national information repositories, simplify access to bureaucratic systems, and enhance 'AI literacy' among the general public. However, the potential of AI for national interest extends beyond these areas, presenting opportunities for targeted efforts in niche domains and education.

Let's first delve into incentivizing AI systems in niche domains. While the development of general-purpose AI has broad applications, there is immense potential for AI solutions tailor-made for specific sectors. For instance, AI in energy could optimize renewable energy production, forecast energy demands, and manage smart grids. In food security, AI could predict crop yields, optimize resource usage, and automate farming processes. In biotechnology, AI could accelerate drug discovery, personalize medical treatments, and decode complex biological systems.

By offering incentives, governments can spur innovation in these critical sectors, fostering national expertise, contributing to domestic well-being, and placing the country on the global map as a leader. These incentives could take various forms, from grants and tax breaks for research in these fields to public-private partnerships that pool resources and expertise.

Another area where governments can foster indirect social and economic benefits is through the establishment of specialized AI learning centers. These centers would not be mere technology institutes but multidisciplinary think tanks that immerse students in AI's technological, ethical, societal, historical, and philosophical aspects.

Such centers would equip students with a deep understanding of AI, producing thought leaders capable of steering the AI revolution in a direction that aligns with societal well-being and national interests. These learning hubs would also attract international talent, creating a virtuous cycle that stimulates further growth and innovation.

These centers could host seminars and public discussions, publish research, and collaborate with international institutions. By doing so, they would foster a culture of openness and knowledge exchange, ensuring the learnings and advancements made in AI are accessible to everyone.

However, establishing such centers and promoting niche AI domains should not be seen as isolated strategies. They are part of a broader national AI strategy that needs to be robust, comprehensive, and flexible. This strategy should outline the nation's AI vision, define the roles of various stakeholders, identify priority areas, and set short-, medium-, and long-term goals.

Governments should seek input from various stakeholders, including academics, industry leaders, civil society groups, and the public when crafting this strategy. This ensures a balanced and inclusive approach, aligning the AI revolution with the aspirations of all citizens.

In the vast ocean of change that the Fourth Industrial Revolution brings, Artificial Intelligence emerges as a colossal wave. Its potential to disrupt and transform is as profound as it is pervasive. As nations sail through these transformative currents, there is a compelling need for a specialized, interdisciplinary approach to harness AI's immense capabilities while mitigating its challenges.

Thus, it is not a question of 'if,' but 'when' we will witness the rise of dedicated "Ministries of AI" in governments around the globe. These ministries, the vanguard of their nations' journey into the future, would serve as nerve centers for coordinating a comprehensive, multifaceted national approach to AI.

Imagine a Ministry of AI orchestrating a symphony of AI applications in public service, streamlining operations, enhancing inter-departmental communication, and empowering citizens through accessible information. Its purview extends beyond implementation, diving deep into the sea of regulations to ensure data protection, combat disinformation, and prevent unethical commercial exploitation.

This Ministry would also be crucial in addressing the elephant in the room - job displacement due to AI. Designing and implementing strategies for reskilling and upskilling the workforce, fostering a culture of lifelong learning, and driving job creation in emerging AI-powered sectors could transform this challenge into an opportunity for economic growth and social progress.

As we stand at the brink of this brave new world, one thing is clear - the future of AI is a tale yet to be written, and it is up to us to write a story that stands the test of time.

Education and public awareness would be among the Ministry's top priorities. Through national campaigns, it would promote AI literacy, dispel fears, debunk myths, and offer a balanced view of AI's potential and pitfalls. This quest would forge alliances with academic institutions, tech companies, non-profits, and international bodies, creating a strong, diverse network that fosters an informed and engaged citizenry.

Furthermore, the Ministry of AI would stimulate innovation by incentivizing research in niche AI domains. It would foster national centers of AI learning, serving as a magnet for global talent and a beacon of knowledge and innovation. Such efforts could place the nation at the forefront of the global AI landscape, all while ensuring the benefits of AI are widely shared and aligned with national interests.

As we look to the future, the emergence of Ministries of AI is not just a futuristic idea but a necessity. Nations that establish such ministries sooner rather than later would be putting their best foot forward in this era of AI-led transformation. The onset of Ministries of AI would signify a proactive recognition by governments of the transformative power of AI. It would represent a critical step towards ensuring that as AI systems continue to evolve, they do so in a manner that is aligned with the broader goals of societal welfare, ethical considerations, and national interests.

AI is a powerful tide that nations cannot afford to ignore. But with vision, collaboration, and a dedicated Ministry of AI, they can not only ride this wave but also steer it in a direction that serves the greater good - a direction that harnesses the power of AI to build a future that is inclusive, equitable, and truly human.

9

WAY AHEAD

As we navigate through this closing chapter, the threads of our narrative begin to weave a vision of the future, where technology and humanity are so deeply intertwined. While basking in the potential and excitement of future possibilities, we should remember to shed light on a few other perspectives to build a comprehensive understanding of the landscape ahead.

Our first stop is the labyrinth of ethical quandaries that AI introduces. These range from questions around content ownership and data utilization to the threat of diminishing incentives for original human creativity.

If an AI generates art or literature, who owns the rights? How do we protect the privacy and rights of individuals when large-scale data collection fuels AI? We will confront These philosophical and practical issues as AI becomes more intertwined with our lives.

Navigating this ethical maze will lead us to the doors of geopolitical and economic consequences. The reshuffling of jobs due to AI and the shift in power dynamics with AI supremacy have far-reaching and transformative implications. Countries and corporations are racing to achieve AI dominance, which could significantly alter the world order. How we handle these shifts will dictate the balance of prosperity and power in our AI-imbued future.

This exciting future with AI is not devoid of shadows. As we cast our collective gaze forward, we must also examine the dangers lurking beneath the surface, the potential misuse of AI, and the profound impact it can have on our society. Just as the power of the atom can be harnessed for energy or devastation, so too can the power of AI. We must recognize these dual realities and create safeguards that minimize harm while maximizing the benefits. As we equip our tools with intelligence, we must also be prepared for the risk that they may be wielded with malintent. It's critical to anticipate these scenarios, not to breed fear, but to prepare defenses and build resilient systems. We cannot afford to ignore this responsibility, for our decisions today will reverberate down the hallways of our future.

Our exploration will only be complete by touching upon our responsibility towards the next generation. We stand on the brink of the AI era, but our children and their children will truly live in it. How do we prepare them for a world where their colleagues could be algorithms, where their creativity will dance with AI's capabilities? This discussion is not just about education and skills; it's about building an understanding, a mindset that will help them thrive in the future.

We will end our journey by contemplating the broader, thought-provoking, existential questions that AI's evolution poses. What is the next leap in our evolution as a species? As we offload more of our tasks to AI and continue to blur the line between man and machine, what does it mean for our understanding of humanity? These are not questions with immediate or definitive answers, but they are the ones that will shape our discourse in the times to come. Though abstract, these questions are essential, making us reevaluate our nature and aspirations.

Discussing these aspects will provoke thought, invite debate, and perhaps unsettle us. Yet, it is this very engagement that is necessary. These are not merely topics for a book; they are the contours of the world we are stepping into.

The Fable of The Digital Eden and The Insatiable Giant

Once upon a time, a vast, unexplored continent emerged on the horizon of our world. It was called the 'Terra-Cyberia,' a boundless expanse where space and time seemed to bend and distort in fascinating ways. In this realm, individuals could sow seeds of their thoughts, ideas, and creativity and share them freely. Over two decades, these seeds, borne out of selflessness and passion, sprouted into a magnificent forest. As more and more people came to marvel at its beauty, some cultivators also discovered they could earn money by inviting visitors to explore their unique patches of digital flora.

The "Terra Cyberia" grew and flourished into a verdant, diverse expanse that stretched as far as the eye could see. The forest was open to all. Anyone who ventured into its depths could freely make copies of any plant, seed, flower, or fruit they encountered, taking a piece of this digital paradise home without any restrictions. But one day, a giant appeared on the outskirts of the forest. He wielded a massive basket with seemingly infinite capacity. With each step he took, the earth shook, and with each movement, he swept up copies of every plant, seed, fruit, and everything else that made up the essence of the forest.

The giant then did something unexpected. He set up a grand restaurant just outside the forest, where he turned the copies of the digital flora into a blend of mouthwatering dishes. Anyone who wanted a taste of his creations could savor a custom-made meal. The giant's dishes were a sensation. Soon, his restaurant was teeming with customers, generating incredible revenue as some patrons were more than willing to pay for the quality of his offerings.

However, this scenario begot an ethical dilemma. The giant had reaped the freely available forest fruits, utilized his resources to process them into delicious dishes, and now profited from the restaurant business. Should he acknowledge every person who had nurtured the forest with their creativity? Given the complexity and depth of the processes involved in his kitchen, was it even feasible?

As his cash registers kept ringing, another question arose. Was the giant morally obligated to share his earnings with those original cultivators whose ideas he had harvested? They had nourished and sustained the forest, fostering its growth without asking anything. If the giant continued his operation unimpeded, would these cultivators still be motivated to tend to their plants and continue to make the digital forest flourish?

The question of the insatiable giant's obligations to the forest and its cultivators gave rise to a spirited debate throughout the "Terra Cyberia." Some inhabitants argued that the giant, through his culinary skills and restaurant initiative, had added value to the ingredients sourced from the forest. His endeavor, risk, kitchen investment, and craftsmanship justified his profit. Moreover, he never broke any rules - the forest was open for all to take as much as they wished.

Others, however, saw the giant's actions as exploitation. They claimed that his venture was only possible with the hard work of the forest cultivators. They argued that his silent harvesting and profiting was a gross injustice to these individuals' efforts. They painted a grim picture of the future, where the forest might wither and die, as the cultivators might decide to stop planting new seeds, seeing no return for their efforts.

At the heart of this ethical problem, the cultivators stood divided. Some continued their work out of pure passion, their joy not in the profit but in the act of creation itself. They believed their work's impact, reflected in the giant's dishes, was reward enough. Others felt cheated, their creativity scooped up by the giant's endless basket, their contribution unacknowledged and uncompensated. They began questioning their incentive to continue working in the forest.

Meanwhile, the giant remained silent, busy catering to the ever-increasing demand in his restaurant. He saw the debate but did not intervene, believing he had done nothing wrong. Yet, deep down, he realized the forest's sustainability was crucial for his venture. With new plants, flowers, fruits, and seeds, his restaurant could continue to surprise its patrons with new flavors.

As the debate continued, multiple suggestions started flowing. Despite his ability to use the forest freely, the giant should contribute to its growth and upkeep. Some proposed a payment system and royalty to the cultivators for their work. The idea was appealing but complex - how to value individual contributions, how to trace ingredients back to their original planters, and how to administer such payments?

Others suggested that the giant reinvest in the forest, funding new tools and resources to help the cultivators plant more diverse and exciting seeds. This seemed more feasible, turning the relationship into a symbiosis where both parties benefit.

While the giant contemplated these ideas, the debate raged, growing more complex daily. The "Terra Cyberia" inhabitants realized this was not just about the giant and the forest cultivators. It was about the very soul of the continent, the principles of its existence, the rules that governed its operation. They understood that whatever path they chose would set a precedent, a model for any future giants who might appear.

The giant's restaurant had become a roaring success, a culinary kingdom in its own right. From the most mundane to the most exotic, the restaurant could whip up any dish the customers desired. The power of the giant's 'infinite kitchen' lay in its vast collection of ingredients gathered from the forest, its uncanny ability to combine these in endlessly creative ways, and its instantaneous response to the whims of the customers. It was, for all intents and purposes, a magical marvel that transformed the act of eating into a spectacle of gustatory delight.

Yet, every success story has an unseen narrative of those pushed into the shadows. The small cultivators who once took pride in their unique creations now struggle against an invisible foe. Their unusual and charming offerings couldn't compete with the boundless variety and instant gratification that the giant's restaurant provided. The footfall in their patches of the forest started dwindling. Their once-thriving spaces were now quieter, the echoes of conversation replaced with an oppressive silence.

A sense of disillusionment crept in among the cultivators. Many found their motivation waning, their joy in creation overshadowed by a sense of impending obsolescence. They started to question the point of planting new seeds, of nurturing novel ideas, when their work was swallowed by the giant's basket, only to be reincarnated in the grand dishes served at his restaurant. A few even gave up, leaving behind their patches of the forest to the relentless march of digital decay.

Yet, amid the despair, some resisted. They rallied the cultivators, reminding them of the original spirit of the Digital Continent – a place for free expression, innovation, and the sharing of ideas. They pointed out that the giant's restaurant, despite its grandeur, was only as good as the forest it relied on. Without their continued cultivation, the restaurant would lose its edge; its menu would become stale and repetitive.

This resistance gave birth to a new movement. A collective of cultivators began experimenting with new types of seeds that would bloom into plants so unique they would defy replication in the giant's kitchen. Others started devising ways to engage with their visitors, offering experiences beyond just viewing and copying, creating a sense of community that a giant restaurant patron might never experience.

At the same time, the cultivators began to voice their concerns more forcefully. They called for a change in how the Digital Continent operated, for rules that would protect their efforts, and for a system to ensure they were rewarded for their creativity. Some even suggested a boycott, a total halt in planting, to make the giant and the continent's inhabitants realize the actual value of their work.

Meanwhile, the giant, watching the stirrings of unrest, couldn't help but feel a tinge of worry. He was reminded, once again, of his dependency on the forest. His 'infinite kitchen' could retain its charm with the continuous growth of unique plants. He was at a crossroads, trying to figure out how to navigate this growing chasm between his booming restaurant and the disgruntled cultivators.

And thus, the saga of the Digital Continent continues, its future hanging in the balance—a tale of discovery, growth, exploitation, and resistance.

The Data Conundrum - IP in The Age of AI

The tale of the "Terra Cyberia" and the Insatiable Giant is not just a fable from a whimsical realm; it's the reality of our digital world, told through metaphor. It beckons us to ponder the ethical implications of harvesting freely available resources for substantial profit without due credit or compensation.

In a conventional setting, the rules of intellectual property are straightforward. A company invests time, resources, and talent in creating technology, subsequently owning the rights to that innovation. It can retain these rights exclusively or license them to other entities for a fee. The cycle is clear: you invest, you innovate, you own. However, this cycle has blurred with the advent of artificial intelligence (AI). Let's take this dilemma of our analogy to the real world - the 'Terra Cyberia' or 'Digital Continent' an expansive forest of data.

The tech companies—the 'Giants'—are skilled chefs. They have mastered the art of cooking up a feast (developing AI algorithms) and opened restaurants (creating products or services). They need ingredients to prepare their dishes (data to train their AI). Instead of growing these ingredients, they walk into the 'Terra Cyberia' and gather whatever they can find.

Here's the catch - this 'Terra Cyberia' was not grown by them. The trees, the fruits, the flowers - all were nurtured by common folks who had no intention of their creations being used as ingredients for someone else's grand feast. They shared their produce freely for others to enjoy, not for it to be harvested and profited from by the Giants.

Now, the question arises - should the Giants be allowed to use these ingredients freely, process them in their kitchen (algorithm), and sell the dishes (products/services) for a profit without compensating the original cultivators? Think of it this way. If a chef opens a restaurant next to a public apple orchard, picks apples for free, makes apple pies, and sells them - we would question its ethics. Shouldn't the public get a share in the profits or, at the very least, be acknowledged?

The Giants are helping 'Terra Cyberia' by creating a demand for its fruits and encouraging more people to contribute. This could be seen as empowering a new generation of creators. However, on the flip side, it also risks undermining the efforts of the original cultivators. It's a delicate balance to strike.

The ethics of this process become even more complex when we consider the cultivators' initial intention. They shared their fruits for human enjoyment, not to be processed and monetized by machine chefs. Can we justify the machine consumption of freely available public data without explicit permission or agreement?

This scenario echoes an ongoing debate in the real world. In today's data-driven economy, tech companies harvest vast amounts of public data, often without explicit consent, to train their AI systems. They then monetize these systems without sharing the profits with the original data creators.

With the emergence of the insatiable giant and his successful restaurant business, others were quick to take note. Like in any gold rush, more big and small giants started trooping into the forest. They began harvesting with abandon, each eager to set up their unique restaurant on the outskirts of the Digital Continent.

Consequently, the Digital Continent, once a haven for dreamers and creators, started transforming. Traffic increased as more giants trudged through the forest with their massive baskets. In response to this predicament, once a free, nurturing ground for anyone to sow and grow their seeds, the farms started thinking about controlling the traffic. They considered imposing a toll on the giants to access their vast expanse. The decision was tough and filled with ethical and practical implications.

The farms, like Twitter, Stack Overflow, and the like, had always been facilitators, providing digital real estate for anyone to grow their ideas. They never interfered with the growth, nor did they take responsibility for the nature of the plants, be they beneficial herbs or poisonous weeds. Could these farms begin charging the giants for their harvest without changing their fundamental nature?

A heated debate erupted among the residents of the Digital Continent. Some argued that charging the giants for access was within the farms' rights. After all, the farms provided the fertile ground on which all these beautiful plants grew, and they were now bearing the brunt of the traffic caused by the giants' activities. They believed the farms deserved compensation for their part in this ecosystem.

Others disagreed. They pointed out that while the farms provided the land, the labor and creativity were those of individual creators. The farms were custodians, not creators. Charging for the harvest felt like claiming rights over someone else's effort. They argued that revenue from such tolls should go to the original creators.

In essence, the rules that were once simple and clear have become complex and murky in the era of AI. Our journey through the digital wilderness reflects the complex interplay between technology, economics, and ethics. As more giants, big and small, descend upon the forest, the need for sustainable and fair practices becomes paramount. The resolution lies not in isolating the giants but in cultivating a shared understanding and responsibility toward our collective digital heritage.

The Digital Continent's evolution mirrored the maturity of human societies. As the giants' actions impacted the forest's tranquility, the farms proposed imposing access fees on these giants as self-defense. However, the implications of such a decision quickly became clear: it could inadvertently birth a monopolistic structure in this vibrant ecosystem, stifling innovation and access.

In this digital gold rush, the early giants who'd already had their fill from the forest could easily afford to pay the new tolls with their vast reserves and revenues. Their restaurants were thriving, the capital at their disposal was abundant, and the price for access was merely a dent in their vast fortunes. They could continue to operate with minimal impact.

However, the situation was starkly different for the new entrants, the startups, and the innovators who had just begun exploring the potential of the Digital Continent to create new restaurants. The tolls could prove prohibitive, turning what was once an open, free space into an exclusive club accessible only to those with deep pockets.

The danger was clear: the vibrant diversity of the Digital Continent could be threatened, with the reins of this vast landscape falling into the hands of a few tech giants. The high entry barriers could thwart creative startups, disruptors, and innovators. This was a classic catch-22 situation, where ensuring the forest's sustainability might lead to a competitive imbalance.

Moreover, the giants, now secure in their positions, might become complacent, no longer needing to strive for innovation or superior service. Their monopoly could choke the very spirit of the Digital Continent, the vibrant churn of creativity, and the free exchange of ideas. The ecosystem's health, once a matter of collective responsibility, could be monopolized by a handful of dominant players.

This problem was not just an economic or technical issue but a moral one. The Digital Continent was a shared heritage, a common ground for all humans to express, innovate, and learn. Should a few determine its destiny? Or should it continue to be the open, inclusive, and diverse space it envisioned?

Thus, the inhabitants of the Digital Continent found themselves at a crossroads. They understood they needed to maintain the forest and manage the traffic without compromising the landscape's inclusive and innovative spirit. The solution was not straightforward, but the dialogue had started. The stakeholders—the farms, the creators, the giants, and the newcomers—were all on this journey together, each playing a crucial role in shaping the Digital Continent's future.

What does the future hold for the Digital Continent? With the growing power of the giants and the changing dynamics of the farms, how do we navigate a fair path for everyone?

Banning the giants from using public data to develop AI solutions is one possibility, but it is fraught with challenges. This could halt the exciting potential of AI, roll back the progress that we have made, and drive these activities underground, making them harder to monitor and regulate. Moreover, finding international consensus in the digitally connected world is arduous. The physical boundaries of nations have blurred in the Digital Continent, making legislation a complex web of jurisdictional dilemmas.

On the other hand, allowing the data providers to charge the AI companies would not be fair either. They merely provide the platform; the content is the labor of millions of individual creators. Tracking and compensating such a large number of people is impossible. It would require a monumental effort in logistics and management, not to mention the privacy and data security issues it would raise.

So, what's the way forward?

One radical idea is to open-source a public information repository. This would create a parallel copy of the forest, accessible only to machines. This approach would solve two problems. It would protect human activity in the original forest from being disrupted and level the playing field for all companies, regardless of their size. Big or small, every giant would have the same access to data.

But this solution also raises its own set of challenges. Who would maintain this new forest? How can we ensure it stays up-to-date with the original? And, importantly, how do we safeguard this vast data repository from misuse or cyberattacks?

As we stand on the brink of this exciting yet challenging future, the role of governments becomes crucial. Until now, they have been spectators, watching from the sidelines as the Digital Continent evolved.

But now, as the situation grows complex, their participation is needed more than ever. It is time they stepped in, studied these emerging dynamics, and initiated a dialogue on the ethical, legal, and practical implications of the Digital Continent.

The journey ahead is uncharted. We must forge our path as we navigate this dense digital forest. But with collective will and proactive governance, we need to ensure that the Digital Continent continues to be a land of opportunity for all, balancing the demands of progress with the principles of fairness and equity.

The Geopolitical and Economic Dominoes

A gentle breeze ripples across the global economy in the quiet before the storm, whispering of the disruption that Artificial Intelligence is about to unleash. As discussed throughout this book, the advancements in generative AI have brought unprecedented efficiencies to the workplace, empowering a single individual to perform tasks that previously demanded entire teams. While teeming with potential, this metamorphosis is also an echo of the approaching thunderstorm of societal disruption.

The storm's effects are evident as we cast our gaze across the horizon of the industrially advanced Western world. Over the past several decades, these nations consciously transitioned from primary and manufacturing sectors to service-oriented economies. The pen of progress has etched a tale of service-oriented economies, with skyscrapers filled with white-collar workers replacing the hum and clatter of factories. The smokestack industries gave way to expensive offices, hard hats were replaced with business suits, and blue collars morphed into white. Creative pursuits were embraced, and domains like Hollywood flourished, broadcasting the seeds of global culture across borders.

A new character has entered the stage - generative AI - and it threatens to disrupt this well-established plot. Automating and innovating offer a potent mix of efficiency and creativity. The result? Armed with AI tools, a single individual can now match the productivity of entire teams. The seemingly unshakeable citadel of white-collar jobs is starting to crumble under the onslaught of generative AI. Industries that once bloomed are now shedding jobs at an alarming pace, their robust structures undermined by the technology they embraced.

As AI grows more competent, we find ourselves standing on the precipice of a paradox. On the one hand, productivity soars. On the other, jobs that were once the economy's lifeblood face a threat of obsolescence. Layoffs, once the province of declining industries, are now making headlines in the very bastions of the future—the tech industry. And this trend, like an unstoppable wave, is set to surge across other service sectors, leaving a wake of job losses.

The rapidity of this disruption poses a daunting challenge. It's a race against time, a sprint to reskill before livelihoods are lost. Yet, the pace of AI's evolution makes it a demanding race many could struggle to keep up with. This imbalance - the yawning gap between job loss and the creation of new roles - could have significant societal implications. Consider the personal strife of those who suddenly find themselves unemployed, the financial strain on families, and the mental stress of uncertainty. Now, multiply these individual stories by thousands, even millions. The societal tapestry, once vibrant with diverse job roles, now risks being frayed by the needle of AI.

Governments, traditionally the safety nets of the unemployed, would find themselves stretched thin. At the same time, the transition could diminish tax revenues as businesses lean more on AI and less on human employees. The strain on public finances would be immense, an economic burden that could redefine policy-making and societal structures.

The Global South: Winds of Change in the IT and BPO Ecosystem

As we shift our focus from the Western world to the Global South, we find ourselves amid a vibrant, bustling hub of Information Technology and Business Process Outsourcing (BPO) services. A critical link in the global supply chain, these economies have blossomed into powerhouses, providing essential services to businesses worldwide, serving as the backstage where the scripts of the global economy are written.

From call centers to code factories, the flourishing sectors of these economies have largely relied on the flow of work outsourced by advanced economies, which found it economically beneficial to tap into the abundant, cost-effective labor in the global south.

However, the winds of change are starting to blow across these landscapes, carrying whispers of AI-driven disruption. A ripple effect, a seismic shift, is anticipated in the burgeoning economies of the global south. The essence of the transformation lies in the rapid advancements in generative AI that are reshaping the coding and data analytics sectors. Low-level programming, data management, and analytics, once the lifeblood of these IT and BPO ecosystems, are now within the striking range of AI tools. A storm brews on the horizon as AI not only automates but elevates these cognitive tasks, executing them with a degree of efficiency and accuracy that humans may struggle to match.

This transformative wave of AI could trigger a massive 'insourcing' trend. Businesses in the West might find it more feasible to use AI tools to execute tasks domestically, achieving greater control over quality, minimizing communication overheads, and perhaps even offsetting the cost advantages of outsourcing.

Visualize a see-saw. On one end, we have the economies of the global south, teeming with IT and BPO services, propelling their economic growth. On the other end, we have the advanced economies of the West, balancing the scale with outsourcing. As AI infiltrates coding and data analytics, the see-saw tilts. The balance shifts towards insourcing as productivity gains offset the cost advantages of outsourcing.

The impending shift of jobs back to the West presents a potential double whammy. Not only does it threaten the loss of existing jobs in the global south, but it also checks the growth of new opportunities. The twin forces of insourcing and AI advancement could destabilize the IT and BPO sectors, casting a shadow of uncertainty over the future of these industries.

The situation could be further compounded if Western governments mandate insourcing to bolster domestic job security. Such policy changes would hit like an aftershock, further shaking the already wobbling infrastructure of the IT and BPO sectors in the global south.

In such a scenario, the global southern economies would grapple with a steep drop in demand. The sprawling IT parks and bustling call centers, symbols of a thriving service economy, could face the specter of downsizing or even closure. This sudden 'insourcing' drive would not just shake the infrastructural foundations of these economies but also cast a pall over the livelihoods of millions who depend on these sectors for their bread and butter.

The Dawn of Reverse Globalization: A Return to Localization

With the tension-laden fabric of geopolitics and the relentless march of AI-driven job disruptions, the crystal ball of the future reveals an intriguing possibility: a shift from an era of burgeoning globalization to a renewed localization. It's a cosmic dance of sociopolitical and economic forces, resulting in an ebb and flow in global dynamics. This dance, performed on the world stage, paints a poignant picture of nations grappling with change, evolution, and survival.

As we probe further into this unfolding future, we find a world map that has begun to transform. Countries in advanced economies and the global south have started to reimagine their economic structures. They draw inward, focusing on building resilient local economies that are buffered from global shocks and less susceptible to volatile international situations. Imagine, if you will, a tapestry of intertwined economies woven together by the threads of globalization. This is now getting strained, tugged at by the forces of power politics and now the disruptive might of AI. As threads fray and knots loosen, countries perceive the need for a sturdy weave of self-reliance. Reflect upon the industrial evolution of the advanced

Western economies, which transitioned from a diverse portfolio of primary, manufacturing, and service sectors to a focused concentration on services. Now, confronted by the seismic political and economic power shifts AI will abet, they're likely to rekindle their latent primary and manufacturing sectors, reviving a balanced trifold economic model.

Picture a nation as an intricate, self-sustaining ecosystem. The primary sectors – farming, fishing, and forestry – form the bedrock, ensuring the nation's food security. The manufacturing sector, the beating heart, generates goods for consumption, jobs, and economic momentum. Lastly, the service sector, the nervous system, provides the essential services that keep the nation humming along. Working in harmony, these three pillars create a sturdy economic architecture that can withstand the tremors of global instability and AI disruption.

The undercurrent of this transformative tide is the necessity to buffer against volatility. In this refocused world, resources become even more precious. Imagine a future where each droplet of water, each lump of coal, and each patch of fertile land takes on increased significance. As populations swell and consumption escalates, these resources might dwindle, becoming points of contention and strategic leverage. Nations may close their fists around their treasures, prioritizing local use and strategic exports.

Take, for instance, a country rich in lithium, a critical element in electric vehicle batteries. In our envisioned future, this nation might restrict its lithium exports, conserving them for local industries or using them as a bargaining chip in international negotiations. Similarly, a nation with vast agricultural lands might prioritize feeding its growing population over exporting food. Take the case of countries rich in rare earth metals like China and Australia. As the demand for these elements in high-tech industries escalates, these nations could resort to strategic export policies, selectively trading these resources for maximum economic or political advantage.

It's a step back, echoing the era before globalization when nations were largely self-sufficient entities. But this new age of localization is not a retreat; instead, it's a strategic adaptation to the complex challenges of our times. It's a testament to nations' resilience and their capacity for transformation and renewal in the face of adversity.

This could be our new normal, a world where every nation, like a well-oiled machine, is self-reliant yet interconnected, operating within a complex web of strategic alliances and partnerships. Nations will start looking at their data more carefully and protecting it, not just the physical, tangible resources.

Harnessing the New Black Gold

Imagine a world before the advent of the automobile. Wide, open prairies, untouched by the tentacles of progress, rested in tranquility, unaware of the 'black gold' they sheltered beneath their serene surfaces. Oil, an unassuming resource, patiently awaited humanity's ingenuity to unearth it and unleash its boundless potential. The advent of the internal combustion engine marked a turning point. This industrial revolution saw oil metamorphose into a commodity of inestimable value.

In a matter of a century, oil's descendants - gasoline, diesel, and jet fuel - power the engines of the global economy, propelling us across land, sea, and sky. The automobile, the airplane, and even the humble plastic—oil advent brought radical transformations that resonate even today. Nations have staked their fortunes on the ebb and flow of oil prices, with economies bolstered or battered by the rhythms of this powerful resource. It turned arid deserts into prosperous city-states and elevated nations to global superpowers. Oil, in essence, became the lifeblood of modern civilization.

Fast-forward to today, and we are in the midst of another extraordinary revolution. This time, it is propelled not by tangible black gold but by the ethereal new oil—data. Just as the mechanical beasts of yesteryear thirsted for petroleum, today's digital giants feed on this invisible resource.

Let's take a moment to ponder: are we witnessing history repeating itself? As we harnessed the potential of oil, are we now on the cusp of fully realizing the vast potential of data?

Now, picture a world before the rise of the digital age. A world untainted by the 24/7 hum of servers, the ceaseless chittering of keyboards, and the unending streams of 1s and 0s crisscrossing the globe. Our lives were not yet intertwined with the powerful algorithmic tendrils of artificial intelligence. Data was just data - raw, unprocessed, and untapped.

Today, we find ourselves on the brink of a new revolution powered not by the underground carbon deposits but by a resource just as valuable and perhaps even more transformative - data. With the rapid evolution of AI technology, we are figuring out ways to extract, process, and refine this new kind of 'digital oil' into forms that can power immense economic engines across various domains.

Just as old refineries transform raw oil into usable fuel, so too do supercomputers and advanced algorithms process vast quantities of data, refining them into invaluable models that can aid us in countless tasks. Like oil, data has been elevated from a byproduct of our digital activities to the driving force behind economic, social, and technological progress.

The emerging narrative draws striking parallels between oil and data. Both are naturally occurring resources that, once harnessed, have the potential to unlock unprecedented economic value. Both require complex infrastructure and sophisticated techniques for extraction and processing.

And, just as nations established sovereignty over their oil reserves, the question arises - should nations protect and monetize their data resources?

However, despite these similarities, the analogy is not perfect. Unlike oil, data isn't bound by geographical territories. It flows freely across national borders through digital streams, making it harder for countries to assert sovereignty over their 'data resources.' Moreover, the value of data lies not just in the sheer quantity but in the patterns and insights that can be extracted through AI and machine learning. It's not just about protecting the raw resource but also the means to refine it.

Reflect for a moment on the process through which oil is formed. Deep beneath the earth's surface, an orchestra of organic material undergoes a slow transformation process. Over the course of millions of years, with the right mix of pressure, temperature, and time, this organic soup is cooked into a hydrocarbon feast – crude oil. Each singular piece of organic matter, in itself, is insignificant. But together, under the right conditions, they form a resource of unimaginable value.

Now, juxtapose this with data. A single data point – an online transaction, a health record, a social media post – is like a single plant in that prehistoric jungle, relatively unremarkable. However, combine millions, billions, and trillions of these data points, and just like those ancient plants under the weight of geological forces, they assume extraordinary value.

Of course, the transformation doesn't stop there. Crude oil must be refined into gasoline, diesel, or jet fuel to be used effectively. Similarly, raw data must be processed, analyzed, and interpreted—refined, if you will—into information and insights that can fuel decision-making, innovation, and progress. This refining process is where artificial intelligence comes into play, providing the techniques and tools to extract meaningful insights from this new resource.

Just as the value of oil is determined by its volume and quality, the wealth of data is gauged by its diversity, quantity, and quality. A vast pool of data, spanning a wide spectrum of categories, brimming with high-quality, reliable information, is akin to a rich oil field, ready to be tapped.

In this context, governments' role becomes pivotal. Governments must recognize the potential wealth within the data generated by their citizens and systems. They need to treat data as they would any other natural resource. This means devising strategies for 'mining' data, setting up the digital infrastructure to facilitate this process, and creating policies that incentivize the production and sharing of high-quality data.

Yet, just as with the extraction and use of oil, there is an ethical and environmental dimension to consider. Data exploitation should not lead to digital divides or breaches of privacy. Principles of equity, respect for privacy, and social good should govern the mining and refining of data. Data should be treated as a common heritage of humankind, promoting prosperity, innovation, and fairness.

We must also consider the risks in our quest to derive value from data. Just as oil spills can cause environmental disasters, data breaches can have serious societal and individual consequences. To protect against such risks, robust cybersecurity measures, effective data governance frameworks, and strong regulations are needed.

As the world barrels headlong into the digital age, nations' wealth is no longer solely in their physical resources but increasingly in the vast, untapped data reservoirs. Recognizing, protecting, and harnessing this wealth is one of the greatest challenges—and opportunities—facing governments today.

A Frightening Deja Vu

An eerie sense of déjà vu permeates the world of technology today. The echoes of the past reverberate ominously in our current trajectory. We may repeat a regrettable chapter of our history if we aren't vigilant. As the idiom goes, those who cannot remember the past are condemned to repeat it.

Picture the late 18th century. The Industrial Revolution was burgeoning, and the world was on the precipice of unprecedented change. Raw, unprocessed natural resources from countries across the globe were plundered and transported, often for no cost, by a select few enterprises to the industrialized nations of the West. There, these resources were transformed by advanced machinery into valuable goods and then sold back at exorbitant prices to the very nations from which the raw materials were taken. The effects of this exploitation were profound and lasting, and the world is still grappling with the repercussions.

Fast-forward to today. We stand at the edge of another precipice, looking out over the vast digital landscape of the 21st century. This time, the resource in question is not iron or coal but data, the new oil of the digital age. Just like the raw materials of the past, it's produced everywhere. Yet, its value is realized only when processed by advanced machinery—in this case, the complex algorithms of artificial intelligence.

A small handful of companies, predominantly in one part of the world, are racing ahead in developing these technologies. They're backed by robust digital infrastructure, deep pockets, and massive data pools, giving them an overwhelming advantage. The parallels to the past are stark and unsettling.

The nations that lag in technology stand to become the digital colonies of these powerful entities, their valuable data exploited. At the same time, they're left to pay the price for the resulting AI products.

In the era of the Industrial Revolution, the exploited nations had little understanding of the true value of their resources and limited means to process them into more valuable goods. Today, many countries similarly lack the understanding of the true worth of their data and the means to harness its value. The result? Vast amounts of data flow freely across borders, feeding into the powerful AI engines of a select few companies.

This trend threatens to create a starkly uneven digital world if left unchecked. A few data-rich companies and nations will become the masters of AI, wielding disproportionate power and influence. Others, dependent on them for AI-driven products and services, risk becoming mere digital colonies — providers of raw data but consumers of finished AI goods.

It's a chilling prospect and a clear call to action for governments worldwide. Recognize the value of your nation's data, and take swift action to protect it. Develop your digital infrastructure and bolster your AI capabilities. Nurture a vibrant ecosystem of AI startups and initiatives to ensure that the benefits of AI are shared equitably and not concentrated in a few hands. The stakes are high, and the urgency is palpable.

This domination will not be just on the objective economic sphere. Still, it will spread further into the subjective and subtle "cultural sphere." Nations will start protecting themselves, furthering the "localization" trend.

Language and Culture: The Silent Frontline in the AI Era

As we trace the intricate mosaic of the future that AI is crafting, we stumble upon an intriguing subplot – the silent tug-of-war between languages and cultures. In a world where data fuels AI, and AI, in turn, molds our interactions with the digital world, language becomes a critical player. In this world,' words' have become as important as 'numbers.' We find ourselves on the cusp of an intriguing dilemma: the preservation of cultural and linguistic identities in the era of AI.

The dominance of English in our globalized world is not just a linguistic phenomenon; it's a cultural, economic, and political narrative that stretches back centuries of history. This linguistic dominance, underpinned by the lingering effects of colonialism and the economic power of the English-speaking world, has spilled over into the digital realm. English has become the 'lingua franca' of the AI world, thanks to the vast trove of digital content it offers. The power of data is as potent as the sword. In the realm of AI, data serves as the lifeblood, fueling and shaping the capabilities of the technology. When it comes to language and culture, the sheer scale of English-language digital content has been a force multiplier, amplifying the predominance of English in AI systems.

Imagine a snowball rolling down a hill, gathering mass and speed as it descends. That's how the English language has grown in the digital landscape. Propelled by the linguistic legacy of colonization and the global south's aspirations to emulate and keep pace with the global north, the digital sphere has become a treasure trove of English data.

Imagine a vast library filled to the brim with books written in English. Each book represents a dataset feeding our AI systems. These systems, trained on English data, inherently perform better when dealing with English content. This creates a self-reinforcing cycle where English-based AI systems proliferate, promoting more English content and tools. As a result, English has become the medium of thought in the digital world, silently asserting its cultural dominance. A self-perpetuating cycle that deepens the digital footprint of the English language and, by extension, amplifies the cultural influence of the English-speaking global north. We are fast approaching a future where English isn't merely a language of communication but a medium of thought in the digital world.

In this spiraling loop, other languages with lesser digital content are pushed to the peripheries. The AI systems trained on these 'minority' languages struggle to perform at par with their English counterparts, leading to a digital divide that mirrors the linguistic divide. This, in turn, can fuel a culture of uniformity, where the monolithic influence of English overshadows diverse linguistic identities.

However, in our imagined future of reverse globalization and localization, nations will not just stand guard over their physical resources but also rise to protect their digital heritage – their languages and cultures. We foresee countries ramping up efforts to digitize their local languages, creating a multilingual digital universe where every language finds its voice.

Having explored the geopolitics of the AI revolution, let's also be aware of "The Looming Storm which will be triggered by the Dangers of AI Misuse."

Dangers of AI Misuse

The Impending Tsunami of Misinformation

In the not-too-distant future, we shall stand at a precipice. Our world, increasingly digitized, will face a new crisis. The machines we've come to rely on, our invisible companions in the great journey of progress, could become our greatest misinformers. In the preceding chapters, we have delved into the dazzling spectrum of this technology's potential. We've marveled at its ability to catalyze change, break barriers, and catalyze unprecedented productivity leaps. And yet, as the future unfolds, the shadowy underbelly of this radiant dawn grows increasingly prominent. The next few pages will explore this darker dimension of our digital destiny, probing the far-reaching dangers that the misuse of AI could unleash.

Despite the occasional intrinsic bias and hallucinations that surface in AI systems, these limitations are far from threats. They are merely inherent shortcomings and glitches in the matrix, which we can confidently predict will be polished and resolved as the technology matures. But herein lies the crux of our dystopian dread: not the AI's weaknesses, but the sinister strength unscrupulous elements could twist into it. The sword of misuse, once unsheathed, can slash at the very foundations of our digital world.

Our deepest fear? A deluge of misinformation. Even at this nascent stage of AI's evolution, we are witnessing an explosion of digital articles crafted not by human hands but by the relentless algorithms of AI. The speed of content creation has accelerated from the patient pace of thoughtful deliberation to the blinding velocity of machine efficiency. Single articles that took hours to compose can now be generated in seconds.

In our relentless quest to democratize AI, we will stumble upon a precarious threshold where the barrier to entry crumbles. A scenario is about to unfold where not just the elite but anyone with a basic understanding of technology will have the power to wield the pen of AI. Prolificacy will skyrocket. The resulting surge of digital articles promises to eclipse traditional human authorship. Yet, the specter of misuse raises its terrifying head amidst this newfound liberty of content creation. In this future, where the traditional barriers to entry crumble into insignificance, any individual, irrespective of intent or credibility, can generate scores of articles in days.

The impending future isn't just about AI revolutionizing our world. It's also about our struggle to navigate the turbulent seas of misinformation, to decipher the convoluted ciphers of truth in an age where the written word can be churned out en masse by using emotionless machines.

In this age of rampant production, the demons of misinformation will dance with impunity. The scale and speed of content creation, empowered by AI, will give rise to a deluge, an unprecedented flood of information that threatens to drown our digital world. The danger is not just in the volume but in the manipulative power of AI. We can imagine this impending reality: AI-authored content's sheer velocity and volume can transform search systems into unwitting accomplices. This future where AI amplifies our voices, we must question – whose voices will be loudest, and will they speak the truth?

Generative AI, like the powerful ocean current, is a marvel of our time. Yet, just as the ocean's waves can erode the shore, this technology can erode our shared truths.

Mass Propaganda - The Silent Puppeteer of Public Opinion

Imagine a world where Generative AI is used as a silent puppeteer, pulling the strings of public sentiment and sewing discord in the tapestry of trust. As we journey into the digital future, state and non-state actors, friend and foe alike, could manipulate this technology to orchestrate a symphony of mass propaganda.

In a not-so-distant future, the state's invisible hand might extend beyond physical borders and use AI to whisper sweet nothings into the ears of global citizens. A government might implant doubt or misinformation about a rival nation. Picture AI creates real-life-looking digital content, a tailored narrative delivered in text, images, videos, or audio. Each item is custom-crafted to cater to individual user preferences, masquerading as innocent media while subtly shaping opinions and sowing distrust.

A single core message, the deceptive song of a digital siren, could be broadcast in many forms. Each rendition is perfectly tailored, like a master key, designed to unlock and mold the minds of its target audience. This could be a slow, almost imperceptible drip of information seeping into the digital landscape over time, acting as a slow poison that eventually taints the well of public sentiment.

On the other hand, non-state actors could harness this technology in a far more ominous manner. Imagine a fringe group using AI to foster unrest, stirring up the embers of discord and fanning them into raging wildfires. They could exploit AI to replicate their message in myriad ways, targeting different demographics and amplifying their reach and influence.

Consider a scenario where AI-generated videos of riots or public disturbances proliferate, inciting fear and anxiety. The nefarious elements could use AI to craft convincing speeches, influential articles, or controversial memes, each tailored to a specific audience, pushing them toward a predetermined viewpoint.

This proliferation of falsehoods, half-truths, and calculated distortions may warp our understanding of reality, blurring the line between fact and fiction, truth and deceit.

These malicious actors could also employ AI to launch sophisticated disinformation campaigns, injecting digital platforms with an insidious cocktail of lies and half-truths. These campaigns could target influential individuals, amplifying the impact and spreading the seeds of their toxic message further and deeper.

In this dystopian future, the menace of mass propaganda will silently slide under our radars, cloaked in the guise of innocuous content. Each piece is a cog in the grand machine of misinformation, turning the gears of public opinion, distorting our perception of reality, and sowing seeds of distrust at an alarming scale.

What we have before us is the terrifying image of inception at scale. In this world, the manipulation of our minds happens so subtly and steadily that we fail to notice. Until, one day, we might wake up to find that our collective consciousness has been steered away from the truth and into the turbulent seas of manipulation and mistrust.

Erosion of Trust - The Silent Siren of a Regression to Analog

Beyond the threats of individual defamation and mass propaganda, an even more insidious consequence of AI misuse lies in its potential to shatter our faith in digital systems. Our reliance on these systems is as profound as it is pervasive, underpinning the entire edifice of our modern, digitized society. Yet, this reliance stands on the precarious ledge of trust, which, if eroded, could send us spiraling backward into an era of analog existence.

Imagine a future where the constant barrage of AI-generated false information turns the digital landscape into a minefield of deception. Each fabricated piece of news, each doctored image, and each twisted narrative are not just independent attacks on truth but coordinated strikes against the integrity of the digital ecosystem.

In the echoing corridors of the future, a chill wind sweeps through, whispering tales of individual, corporate, and institutional reputations tarnished by AI-enabled slander.

This misuse of Generative AI can turn it into a weapon of digital defamation, capable of assaulting the integrity of anyone or anything in the crosshairs. The future might witness a highly respected CEO brought low by a swarm of AI-generated articles spreading rumors about malpractice. Or imagine a renowned institution's reputation stained by a flurry of synthetic videos alluding to fabricated scandals. Even a seemingly untouchable corporation could see its stock prices plummet as AI creates and propagates false financial reports, prompting an unwarranted panic among investors.

As we erect defenses to detect and contain such fake news, we must confront a chilling reality: the damage is often done before our digital shields can rise. By the time we debunk a lie, it could have already traveled halfway around the world, nestling into the minds of millions, painting the accused in shades of mistrust. This perpetual game of cat-and-mouse with misinformation will harm individuals or institutions and gnaw at the foundations of our faith in digital systems. The onslaught of genuine-looking fake news could easily outpace the truth, prompting us to question everything we see, hear, or read online.

Suppose the uncertainty and fear of being deceived start to overshadow the convenience and accessibility of digital sources. In that case, the pendulum of preference may swing back towards the tangible certitude of paper and books. Once a beacon of knowledge and a testament to humanity's progress, the digital library could transform into a haunted house of falsehoods, driving users away in droves.

The convenience of a tap or a click could lose charm when juxtaposed with the fear of deceit. As our trust in digital sources withers, we might yearn for the perceived security of paper and books, craving the tactile authenticity they promise.

This prospect of regression to analog, driven by a collective disillusionment with the digital world, casts a shadow over our march toward an information-rich future. The repercussions of such a transition would be seismic. We have come to rely heavily on the efficiency, accessibility, and scalability of digital information. It powers our global economy, communication systems, knowledge repositories, and more. The process of digitization, which has drastically transformed our information landscape, stands at the risk of being undone, pushing us back into an era of physical records, slower dissemination, and constrained accessibility. Will we forsake the tremendous strides we have made in digitizing and organizing our information, retreating to the confines of tangible records out of fear and mistrust?

Information Bubbles and the Echo Chamber Effect - A Polarized Society

As we wade deeper into the potential dangers of AI misuse, we stumble upon another disturbing possibility - the amplification of information bubbles. In our current digital ecosystem, sophisticated interest-tracking systems curate our information diet, serving us with content that aligns with our preferences and beliefs. In the vortex of this self-reinforcing echo chamber, our inherent confirmation bias is further fueled, resulting in a dangerous cycle that's primed to spiral out of control with the advent of generative AI systems.

Picture a future where this technology is unleashed onto our personalized digital landscapes. Like an overzealous gardener, it meticulously plants and nurtures only those ideas that align with our pre-existing beliefs, effectively starving and stifling dissenting viewpoints. With each scroll and click, we unknowingly draw deeper into the labyrinth of our convictions, insulated from the healthy challenge of differing perspectives.

In this not-too-distant dystopia, your morning coffee is accompanied by AI-generated news articles that reinforce your worldview and avoid contrasting opinions. As your day unfolds, AI-curated social media feeds and video suggestions continue to echo your beliefs, meticulously editing out any discordant notes.

The consequences of this reinforcement will ripple across all layers of society, polarizing opinions on topics ranging from mundane personal preferences to critical national decisions. For instance, imagine a local community split over a decision to install wind turbines. One group hails it as a step towards sustainable energy; the other views it as a noisy eyesore. As AI feeds each group's biased information, the rift between them could widen, and what could have been a healthy debate becomes a deep-seated division.

On a national scale, such polarization could disrupt policy-making, elections, and even the nation's social fabric. Imagine a country divided over its immigration policy. With AI systems amplifying their respective beliefs, the proponents and opponents become entrenched in their positions, making compromise or consensus increasingly elusive.

This AI-augmented polarization threatens to create a world where understanding and compromise give way to intolerance and division. A world where decision-making, whether at the personal, community, or national level, becomes a battleground rather than a negotiation table.

In this future, the echo chambers we live in will become fortified fortresses, and the bridge of balanced discourse will crumble under the weight of reinforced bias.

The Real AI Apocalypse - An Implosion Rather than an Explosion

When we conjure images of an AI-induced apocalypse, our imagination, fuelled by the relentless portrayal in Hollywood and popular media, typically travels towards a world reminiscent of "The Terminator" or "The Matrix." A dystopian future where cold, ruthless machines seize control of humanity and the planet, a reign of technological terror. While such a scenario captivates our collective imagination, the real threat of AI misuse is much more subtle, insidious, and unnervingly imminent.

The true AI apocalypse may not come with the sound of gunfire or the cold gleam of a robotic overlord. Instead, it may arrive quietly, gradually, almost imperceptibly, manifested in the slow implosion of our society. A silent siege led by the unholy trinity of subtle propaganda amplified information bubbles and job displacement due to rapid AI adoption. Rather than an external invasion by sentient machines, the real AI apocalypse may arise from an internal implosion of society spurred on by a myriad of factors related to the misuse of AI. This could be a more subtle and gradual process, but its impact could be just as devastating, if not more so.

Consider the influence of propaganda. This tool has been used throughout history to control and sway public opinion. But imagine this weapon supercharged by AI's capabilities. The propaganda machine, bolstered by advanced AI, would generate real-life-looking content tailored to each individual's preferences and worldviews, subtly shifting opinions and insidiously sowing seeds of discord. The slow drip of altered facts and skewed perspectives could eventually alter the course of elections, induce social unrest, or even trigger international conflicts.

This manipulation could be so nuanced and relentless that we might not even recognize the inception of ideas that aren't our own. Compounding this, we face the amplification of information bubbles. As AI becomes more adept at tailoring content to our individual preferences, we could find ourselves trapped within echo chambers that continuously reinforce our existing biases and beliefs. The resulting polarization could lead to societal fragmentation, where finding common ground becomes increasingly difficult. Simultaneously, the rapid transition of jobs due to AI adoption could lead to widespread social and economic upheaval. As AI technologies mature and permeate various industries, we could witness a massive shift in the job market. Traditional roles could disappear overnight, replaced by jobs requiring a whole new skill set. Those unable to adapt quickly could find themselves in a precarious situation, amplifying economic disparities and societal tension. Without adequate measures to manage this transition, we could witness unemployment, a surge in social inequality, and a profound sense of displacement among those whose skills are no longer required.

Picture a society grappling with widespread job displacement. At the same time, its members are increasingly locked within their echo chambers, their worldviews distorted by subtle propaganda. The tensions between different societal segments could escalate as understanding and empathy are eroded by isolation and misinformation. This pressure-cooker situation, where society is simultaneously subjected to economic, informational, and social stresses, could lead to a societal implosion, a collapse from within.

The potential for societal disruption posed by the misuse of AI is not some distant threat; it's a real possibility looming. While robot-led apocalypses make for good cinema, it's these subtler threats that we need to be vigilant about. These factors, individually and collectively, represent the dangers that AI misuse poses to our society and underscore the urgent need for thoughtful and comprehensive governance of AI technologies.

No magic bullet or simple solution could effortlessly extinguish the challenges that lie before us in navigating the complexities of potential AI misuse. However, one thing remains abundantly clear: we cannot afford to ignore these potential threats. Turning a blind eye is not an option when the stakes are this high.

Historically, technological advancements have been shaped predominantly by economic motives and private entities. However, considering the profound societal impact AI can have, this approach is no longer viable. The stewardship of AI cannot be left solely to the whims of market forces or the strategic interests of private corporations. The path to mitigating the misuse of AI and harnessing its potential for societal good lies in collective action.

The evolution of AI must be a comprehensive narrative encompassing all facets of society. Governments, thought leaders, philosophers, civil rights groups, investors, and enterprises uniquely shape this dialogue; each voice is essential.

Governments can regulate and create a policy framework that safeguards public interest and mitigates potential misuse. Thought leaders and philosophers can shape the ethical discourse around AI, ensuring that the development and deployment of these technologies align with our shared values. Civil rights groups can act as watchdogs, advocating for the rights and interests of citizens and keeping the powers that be accountable. Investors and enterprises must continue to fund innovation responsibly while emphasizing ethical AI practices. The future of AI is too important to be left to chance or driven solely by economic motives.

In the midst of the Second Renaissance, we find ourselves facing the rise of a new kind of intelligence - artificial yet boundlessly generative. It is a fascinating time, akin to witnessing the first sparks of fire or the inaugural flight of the Wright brothers. As we marvel at the innovative leaps and bounds, we must realize that our actions, choices, and inventions will shape the fabric of our descendants' reality. The stakes couldn't be higher - and the onus of responsibility is, as such, tremendous. Amidst this responsibility, educating our children and the next generation has never been more critical. Our children are, in fact, at the center of this immense transformation. They will inherit our planet, our choices, technology, and consequences. Suppose our actions echo the depths of time. In that case, our children are the receivers, and their lives are shaped by the repercussions of our decisions today. There is an urgency to prepare them to equip them with the knowledge and wisdom they will need in the age of AI. We cannot neglect or defer this responsibility to future generations, given that the ripple effects of our present deeds, discoveries, and choices will reverberate far beyond our fleeting existence.

Empowering Our Next Generation for the Age of AI

In this context, the term "AI literacy" gains unprecedented importance. However, it is not merely about coding skills or understanding neural networks. AI literacy transcends technical knowledge, delving into understanding its societal implications and potential and having the maturity to wield this potent tool effectively. It is about empowering our children to harness AI and become charioteers controlling its speed and direction rather than mere passengers borne by its tide. They need to learn how to separate the wheat from the chaff - to distinguish between what AI can and should do and what it can but shouldn't.

We are at the precipice of an era where our creations may surpass us in capabilities we've held as uniquely human. Let's consider the evolution of mathematical skills. Once, the quickest mental calculators were deemed indispensable. Today, however, knowing what calculations to perform takes precedence over speed - a shift necessitated by machines relieving us of the numerical heavy lifting. We foresee a similar transition for language skills in the not-so-distant future. Superior language fluency may soon cease to be the gauge of expertise on a topic. Instead, the ability to think clearly and communicate effectively, regardless of verbosity, will take the helm. These will be the fundamental skills - the linchpin of our children's intellectual repertoire.

Superior fluency should not be mistaken for mastery of a subject. Effective thinking and communication must become fundamental skills, for in the era of AI, language will transcend being a mere tool of expression to become an instrument of control and understanding. In the AI-driven future, conversing fluently with AI, questioning it, learning from it, and even challenging it will become essential skills. Thus, our education must imbue our children with critical thinking, logical reasoning, curiosity, and skepticism, even as they learn to live, work, and create alongside AI.

It is about teaching them the art of asking the right questions rather than always having the right answers.

Our children must be more than passive consumers of AI; they must be its architects and controllers. Education must shift from a traditional approach focused on rote memorization and regurgitation to an inclusive model that promotes original thinking, creativity, ethical awareness, and responsible AI usage. This is not an optional upgrade to our education system but a crucial transformation, as necessary as the shift from parchment to digital screens.

Undoubtedly, the skill set required to thrive in an AI-influenced job market or economy will evolve rapidly. But let's not forget - the essence of human intelligence lies not just in our ability to acquire and apply knowledge but in our adaptability. Therefore, we must equip the next generation with the tools of resilience, agility, and the wisdom to question, explore, and lead. Our collective aim should not merely coexist with AI but shape its development, ensure its ethical use, and harness its potential to solve humanity's most pressing problems.

As we envision the age of generative AI, we must acknowledge that our existing education system, largely a product of the industrial age, may not be equipped to address the needs of a society embedded in an AI-driven reality. We are facing a new dawn, an era where information is the new currency, and learning how to manage it effectively is vital.

This information age requires citizens capable of navigating an ever-expanding digital universe. Our children will not just passively consume information but actively engage with a tsunami of data, a constant deluge of hyper-personalized digital content. They must learn how to filter, analyze, interpret, and synthesize this information, transforming it from mere data into actionable knowledge.

Therefore, the 'reading, writing, and arithmetic' of the past must evolve into 'researching, reasoning, and reflecting' in the present. The ability to ask insightful questions, seek out reliable sources, and distill truth from falsehood while thinking critically and creatively should form the bedrock of this new education.

Moreover, as we embed AI deeper into our lives, we must guard against its misuse. One of the most daunting challenges our children will face is the distortion of truth, the rise of deep fakes, and AI-driven misinformation. Therefore, we must arm them with the ability to discern fact from fiction, see through the smoke and mirrors of digital illusion, and pursue truth despite the biases that may color their information landscape.

Critical media literacy, therefore, will be an essential skill. Our children need to understand not just what the media is telling them but why it's telling them, who's behind the message, and how it's being presented. We must instill a healthy skepticism towards the information they encounter, encouraging them to probe, question, and validate before accepting it as truth.

The AI-integrated society demands a more conscious, active approach to learning and understanding. It's a shift from absorbing to questioning, from passivity to active engagement. The education of the future should empower our children not only to thrive in this new world but to shape it to mold it in the image of the just, ethical, and inclusive society we aspire to create.

In forging the educational blueprint of the future, we must remember that the technology our children will wield is as capable of mirroring our prejudices as it is our potential. A primary aspect of this new educational paradigm must be a heightened awareness of biases – both our own and those inherent in our sources of information. We must teach our children that no source of information is completely neutral and each carries its prejudices. Recognizing these biases allows us to view the information through a lens of discernment. It enables a more balanced and nuanced understanding of the world.

In an age where algorithms increasingly shape our experiences, our children must also be aware of the pitfalls of information bubbles. These bubbles reiterate and reinforce our existing views and can expose us to alternate perspectives and ideas. Our education must encourage exploration and curiosity, urging our children to venture beyond the comfort of the familiar and actively seek out diverse opinions, experiences, and knowledge. We must teach our children not to surrender to the seductive gravitational pull of these bubbles but to traverse beyond, explore different perspectives, and understand and respect the diversity of thought.

The real world, too, must not become a casualty of the digital age. While the digital realm offers countless opportunities, it is crucial to balance virtual experiences with the tangible realities of the physical world. As the lines between the physical and the virtual blur, we must ensure our children understand the importance of human values, empathy, compassion, cooperation, and social responsibility. We must foster in them a sense of global citizenship that transcends digital borders, urging them to apply their skills and knowledge to address pressing human issues. It's not enough to prepare them for the world; we must inspire them to envision it as it could be.

Let's prepare the road for them and prepare them for the road ahead!

We must recognize that transforming the education system is a marathon, not a sprint. We must recognize that transforming the education system is slow, hindered by many stakeholders, convoluted decision chains, and invariable time. Yet, we cannot afford to wait. The responsibility to instill "AI literacy" starts at home. As parents, guardians, caregivers, or mentors, we must first equip ourselves with the necessary understanding of AI. Through our informed conversations, engaging discussions, and shared explorations, we can lay the foundation for our children's understanding of AI. In these daily dialogues, we can spark their curiosity, foster critical thinking, and guide their moral compass.

In this era of rapid and exponential tides of change, we cannot abdicate our responsibility and cry helplessness in the face of change. If we do not hold the reins today, our children might get caught in a vortex tomorrow. Instead, we should stand as their lighthouses, illuminating their paths with wisdom. Our homes should become the first classrooms where they learn to navigate change currents and question, think, and grow.

The Metamorphosis of Humanity in the AI Era

An Ode to Our Future

As we reach the grand finale of our sweeping saga through the evolution of AI and its future, we're led to a point of introspection. At this juncture, we must reflect on our journey as humans in the ever-unfolding drama of technological innovation. We must look inward at the fabric of our existence and transformations, for it's the essence of who we are that shapes and is shaped by the sweeping waves of technology. Through our journey, we've been historians, observers, and explorers of AI. Still, it's time to be philosophers and visionaries, for the dawn of AI is not merely about technological shifts but a transformative era that could redefine humanity itself.

Akin to a mirror reflecting our image, our exploration of technology reveals an interesting pattern. From the advent of the steam engine to the rise of industrial machines, technology has primarily served as an extension of our physical capabilities, enabling us to conquer distances, time, and scale in ways unimaginable to our ancestors. It's fascinating, almost paradoxical, how our physical strength as a species has dwindled in the same proportion as our technological prowess in the physical domain has skyrocketed.

Take a moment to envision the laborers of ancient Egypt, their sinewy muscles straining against the colossal weight of the stones that would form the Great Pyramids. Consider the knights of the Middle Ages; their physical strength was tested to the limit as they hoisted hefty swords and shields, their bodies encased in heavy armor. Today, such feats of physical strength are rare, almost anachronistic, in our mechanized world. Yet, that same individual, armed with the knowledge and cognitive capabilities powered by today's education and technology, can command these machines to perform tasks that would have once seemed impossible. As a species, it's as if we made a subconscious pact to trade physical might for mental acuity.

Machines, an offspring of our inventive minds, replaced our physical tasks. In this metamorphosis, we found ourselves evolving, shifting our focus from the strength of our bodies to the power of our minds. A touch on a screen and a click of a mouse are the tools of our time, a testament to the cerebral nature of our existence in the 21st century. We are staring at a future where AI and machine learning are becoming increasingly adept at mental tasks, and the implications of this development are profound. Will this trend weaken our mental acuity, mirroring the decline in our physical strength? Or will it catalyze the emergence of novel faculties and capabilities, the next step in our evolution as a species?

Imagine a scale and its two sides weighed down by these vastly different outcomes. On one side lies the possibility of cognitive decline, a chilling echo of our diminished physical prowess. Conversely, a more hopeful prospect gleams—the birth of new dimensions of human consciousness and creativity fueled by our symbiotic relationship with AI.

Indeed, both scenarios seem plausible, as their seeds are already sown in the present. The choice we make now—whether to foster a harmonious symbiosis with AI or let the consequence of our inventions wreak havoc on our social structures—will define our future.

The coin is in the air, spinning on the axis of time. Will we let the coin drop at the mercy of chance, or shall we, with thoughtful intent and a shared vision, ensure a toss that lands favorably for all of humanity?

The choice is ours.

ABOUT THE AUTHOR

Aswin Chandarr is a distinguished figure at the forefront of artificial intelligence (AI) and robotics, embodying the roles of an entrepreneur, author, and educator. With over 18 years of ground breaking work in AI and robotics, Chandarr has dedicated his career to unraveling the complexities of technology and its implications for humanity's future.

As the visionary founder of The Global AI Transformation Institute, Chandarr has created a pivotal platform committed to demystifying AI's vast capabilities and potential for societal benefit. Promoting a multidimensional view of technology as a force for global progress and innovation, it offers unparalleled consulting and training that intersects business, education, and policy.

With a rich academic background, including a Ph.D. and an MBA from prestigious institutions in the Netherlands and the USA, Aswin can demystify complex technical terms with astonishing simplicity. His work extends beyond academia into the tangible world, where he has pioneered multiple robotics companies, such as Robot Care Systems and Loop Robots. These ventures focus on integrating robots into daily life to address some of society's most pressing challenges, particularly in elderly care and healthcare.

Chandarr's work as a futurist is not just about predicting the future; it's about creating it. His insights into the evolving landscape of AI and robotics are profound and actionable, encouraging others to engage proactively with the future of technology.

For further details, deeper insights, or to connect with the author for inquiries, feedback, or collaboration opportunities, simply scan the QR code.